Global Warming

Global Warming

Edited by
Declan Hernandez

Larsen & Keller
www.larsen-keller.com

Global Warming
Edited by Declan Hernandez
ISBN: 978-1-63549-138-8 (Hardback)

▤ Larsen & Keller

Published by Larsen and Keller Education,
5 Penn Plaza,
19th Floor,
New York, NY 10001, USA

Cataloging-in-Publication Data

Global Warming / edited by Declan Hernandez.
   p. cm.
Includes bibliographical references and index.
ISBN 978-1-63549-138-8
1. Global warming. 2. Greenhouse effect, Atmospheric. 3. Climatic changes.
I. Hernandez, Declan.
QC981.8.G56 G56 2017
363.738 74--dc23

This book contains information obtained from authentic and highly regarded sources. All chapters are published with permission under the Creative Commons Attribution Share Alike License or equivalent. A wide variety of references are listed. Permissions and sources are indicated; for detailed attributions, please refer to the permissions page. Reasonable efforts have been made to publish reliable data and information, but the authors, editors and publisher cannot assume any responsibility for the vailidity of all materials or the consequences of their use.

Trademark Notice: All trademarks used herein are the property of their respective owners. The use of any trademark in this text does not vest in the author or publisher any trademark ownership rights in such trademarks, nor does the use of such trademarks imply any affiliation with or endorsement of this book by such owners.

The publisher's policy is to use permanent paper from mills that operate a sustainable forestry policy. Furthermore, the publisher ensures that the text paper and cover boards used have met acceptable environmental accreditation standards.

Printed and bound in the United States of America.

For more information regarding Larsen and Keller Education and its products, please visit the publisher's website www.larsen-keller.com

Table of Contents

Preface

Global Warming refers to the increase in the earth's total temperature, which results in melting of glaciers and icebergs, thus, increasing the water level by alarming rates. Global Warming is seen to be single-handedly damaging most of our environmental resources. It also leads to depletion of ozone layer, which results in increasing temperature and pollution. This textbook on global warming talks about the fundamental concepts that redefine this field. The aim of this book is to provide the students with detailed information about the primary concepts of global warming. It includes the topics of utmost importance which are bound to provide incredible insights to the readers. Some of the diverse points covered in this text address the varied branches that fall under this category. This textbook will serve as a valuable source of reference for those interested in this field.

To facilitate a deeper understanding of the contents of this book a short introduction of every chapter is written below:

Chapter 1- The Earth's climate is in serious crisis with the emission of greenhouse gases into the ozone layer. Reports estimate that with the dramatic decrease of polar ice, the ability of the Earth's surface to reflect solar radiation has also dwindled. This chapter is an overview of the subject matter incorporating all the major aspects of global warming.

Chapter 2- It can be proven outright that the exploitation of the Earth's natural resources by human activity has led to uncontrollable levels of greenhouse gas emissions. These emissions have to be systematically shut down in order to mitigate climate change. This chapter lists the major causes of global warming and how they are interconnected with each other.

Chapter 3- The atmosphere is a mixture of different gases that facilitate chemical and physical activity. But a sharp increase or decline in any one of these gases during a short interval can cause imbalances in the atmosphere and the surface. The chapter strategically encompasses and incorporates the major components and key concepts of greenhouse gases, providing a complete understanding.

Chapter 4- Climate change can cause significant damage to humans, the environment and animals. All living things depend on a fine balance of natural interventions, such as high levels of rainfall or drought as well as predictable, seasonal and hospitable climates. Global warming can alter these much-desired phenomena for the worse. Global warming is best understood in confluence with the major topics listed in the following chapter.

I owe the completion of this book to the never-ending support of my family, who supported me throughout the project.

Editor

Introduction to Global Warming

The Earth's climate is in serious crisis with the emission of greenhouse gases into the ozone layer. Reports estimate that with the dramatic decrease of polar ice, the ability of the Earth's surface to reflect solar radiation has also dwindled. This chapter is an overview of the subject matter incorporating all the major aspects of global warming.

Global warming and climate change are terms for the observed century-scale rise in the average temperature of the Earth's climate system and its related effects. Multiple lines of scientific evidence show that the climate system is warming. Although the increase of near-surface atmospheric temperature is the measure of global warming often reported in the popular press, most of the additional energy stored in the climate system since 1970 has gone into ocean warming. The remainder has melted ice and warmed the continents and atmosphere.Many of the observed changes since the 1950s are unprecedented over tens to thousands of years.

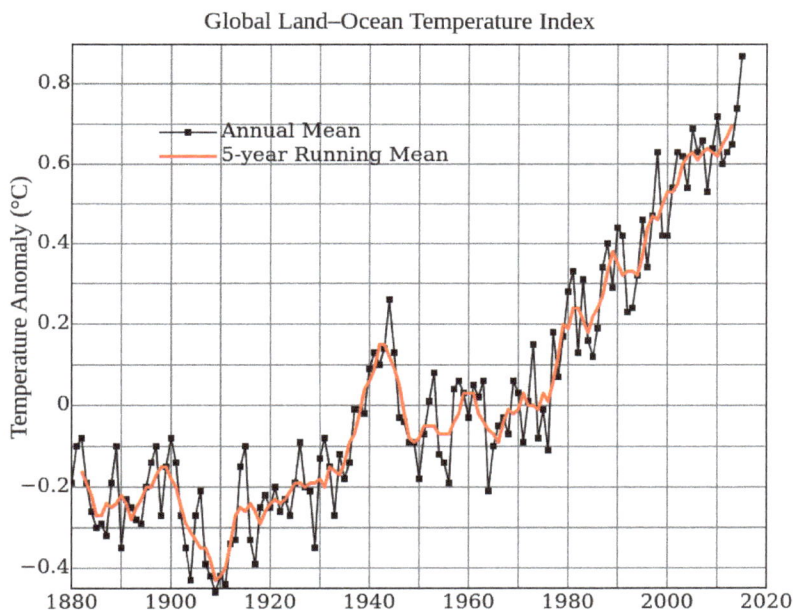

Global mean surface temperature change from 1880 to 2015, relative to the 1951–1980 mean. The black line is the annual mean and the red line is the 5-year running mean. Source: NASA GISS.

Scientific understanding of global warming is increasing. The Intergovernmental Panel on Climate Change (IPCC) reported in 2014 that scientists were more than 95% certain that global warming is mostly being caused by human (anthropogenic) activities, mainly increasing concentrations of greenhouse gases such as carbon dioxide (CO_2). Human-made carbon dioxide continues to increase above levels not seen in hundreds of thousands of years: currently, about half of the carbon dioxide released from the burning of fossil fuels is not absorbed by vegetation and the oceans and remains in the atmosphere. Climate model projections summarized in the report indicated that during the

21st century the global surface temperature is likely to rise a further 0.3 to 1.7 °C (0.5 to 3.1 °F) for their lowest emissions scenario using stringent mitigation and 2.6 to 4.8 °C (4.7 to 8.6 °F) for their highest. These findings have been recognized by the national science academies of the major industrialized nations and are not disputed by any scientific body of national or international standing.

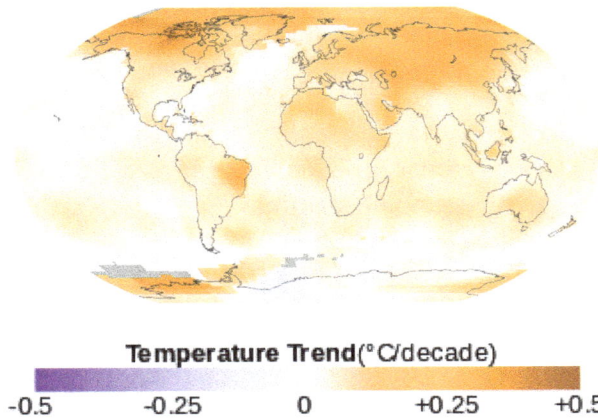

World map showing surface temperature trends (°C per decade) between 1950 and 2014. Source: NASA GISS.

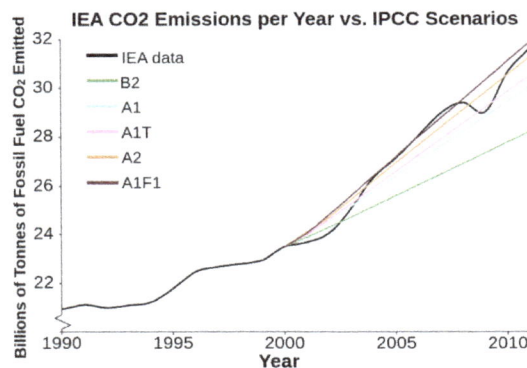

Fossil fuel related carbon dioxide (CO_2) emissions compared to five of the IPCC's "SRES" emissions scenarios, published in 2000. The dips are related to global recessions. Image source: Skeptical Science.

Fossil fuel related carbon dioxide emissions over the 20th century. Image source: EPA.

Future climate change and associated impacts will differ from region to region around the globe. Anticipated effects include warming global temperature, rising sea levels, changing precipitation, and expansion of deserts in the subtropics. Warming is expected to be greater over land than over the oceans and greatest in the Arctic, with the continuing retreat of glaciers, permafrost and sea ice. Other likely changes include more frequent extreme weather events including heat waves, droughts, heavy rainfall with floods and heavy snowfall; ocean acidification; and species extinctions due to shifting temperature regimes. Effects significant to humans include the threat to food security from decreasing crop yields and the abandonment of populated areas due to rising sea levels. Because the climate system has a large "inertia" and CO_2 will stay in the atmosphere for a long time, many of these effects will not only exist for decades or centuries, but will persist for tens of thousands of years.

Possible societal responses to global warming include mitigation by emissions reduction, adaptation to its effects, building systems resilient to its effects, and possible future climate engineering. Most countries are parties to the United Nations Framework Convention on Climate Change (UNFCCC), whose ultimate objective is to prevent dangerous anthropogenic climate change. The UNFCCC have adopted a range of policies designed to reduce greenhouse gas emissions and to assist in adaptation to global warming. Parties to the UNFCCC had agreed that deep cuts in emissions are required and as first target the future global warming should be limited to below 2.0 °C (3.6 °F) relative to the pre-industrial level, [c] while the Paris Agreement of 2015 stated that the parties will also "pursue efforts to" limit the temperature increase to 1.5 °F (0.8 °C).

Public reactions to global warming and general fears of its effects are also steadily on the rise, with a global 2015 Pew Research Center report showing a median of 54% who consider it "a very serious problem". There are, however, significant regional differences. Notably, Americans and Chinese, whose economies are responsible for the greatest annual CO2 emissions, are among the least concerned.

Observed Temperature Changes

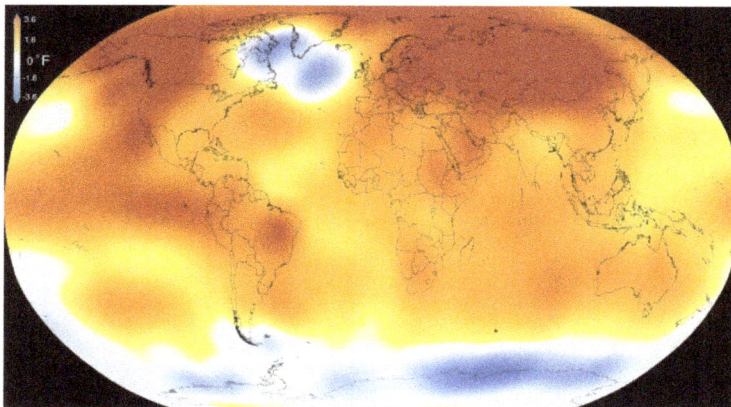

2015 – Warmest Global Year on Record (since 1880) – Colors indicate temperature anomalies (NASA/NOAA; 20 January 2016).

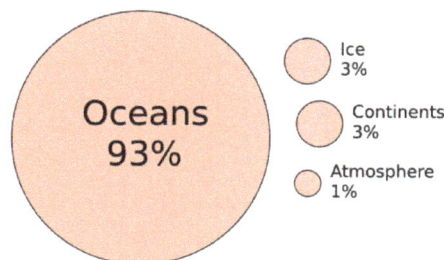

Energy change inventory, 1971-201(

Oceans 93%

Ice 3%

Continents 3%

Atmosphere 1%

Earth has been in *radiative imbalance* since at least the 1970s, where less energy leaves the atmosphere than enters it. Most of this extra energy has been absorbed by the oceans. It is very likely that human activities substantially contributed to this increase in ocean heat content.

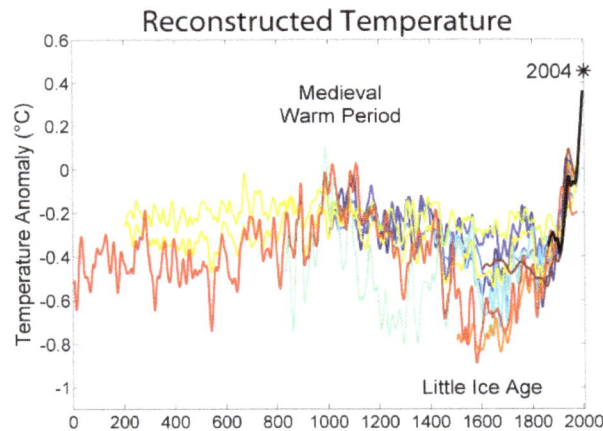

Two millennia of mean surface temperatures according to different reconstructions from climate proxies, each smoothed on a decadal scale, with the instrumental temperature record overlaid in black.

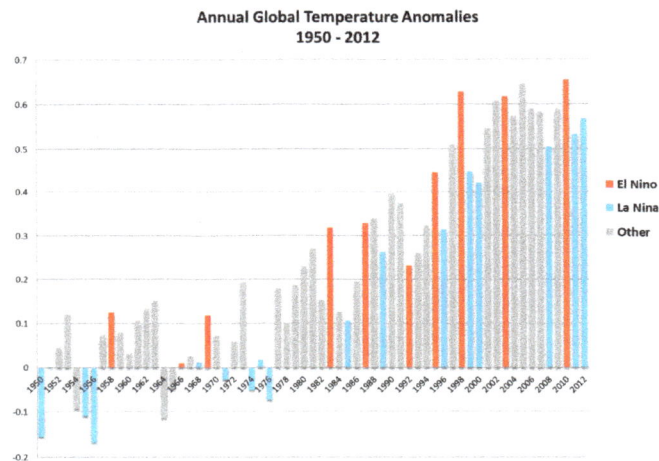

NOAA graph of Global Annual Temperature Anomalies 1950–2012, showing the El Niño Southern Oscillation

The global average (land and ocean) surface temperature shows a warming of 0.85 [0.65 to 1.06] °C in the period 1880 to 2012, based on multiple independently produced datasets. Earth's average surface temperature rose by 0.74±0.18 °C over the period 1906–2005. The rate of warming almost doubled for the last half of that period (0.13±0.03 °C per decade, versus 0.07±0.02 °C per decade).

The average temperature of the lower troposphere has increased between 0.13 and 0.22 °C (0.23 and 0.40 °F) per decade since 1979, according to satellite temperature measurements. Climate proxies show the temperature to have been relatively stable over the one or two thousand years before 1850, with regionally varying fluctuations such as the Medieval Warm Period and the Little Ice Age.

The warming that is evident in the instrumental temperature record is consistent with a wide range of observations, as documented by many independent scientific groups. Examples include sea level rise, widespread melting of snow and land ice, increased heat content of the oceans, increased humidity, and the earlier timing of spring events, e.g., the flowering of plants. The probability that these changes could have occurred by chance is virtually zero.

Trends

Temperature changes vary over the globe. Since 1979, land temperatures have increased about twice as fast as ocean temperatures (0.25 °C per decade against 0.13 °C per decade). Ocean temperatures increase more slowly than land temperatures because of the larger effective heat capacity of the oceans and because the ocean loses more heat by evaporation. Since the beginning of industrialisation the temperature difference between the hemispheres has increased due to melting of sea ice and snow in the North. Average arctic temperatures have been increasing at almost twice the rate of the rest of the world in the past 100 years; however arctic temperatures are also highly variable. Although more greenhouse gases are emitted in the Northern than Southern Hemisphere this does not contribute to the difference in warming because the major greenhouse gases persist long enough to mix between hemispheres.

The thermal inertia of the oceans and slow responses of other indirect effects mean that climate can take centuries or longer to adjust to changes in forcing. Climate commitment studies indicate that even if greenhouse gases were stabilized at year 2000 levels, a further warming of about 0.5 °C (0.9 °F) would still occur.

Global temperature is subject to short-term fluctuations that overlay long-term trends and can temporarily mask them. The relative stability in surface temperature from 2002 to 2009, which has been dubbed the global warming hiatus by the media and some scientists, is consistent with such an episode. 2015 updates to account for differing methods of measuring ocean surface temperature measurements show a positive trend over the recent decade.

Warmest Years

15 of the top 16 warmest years have occurred since 2000. While record-breaking years can attract considerable public interest, individual years are less significant than the overall trend. So some climatologists have criticized the attention that the popular press gives to "warmest year" statistics; for example, Gavin Schmidt stated "the long-term trends or the expected sequence of records are far more important than whether any single year is a record or not."

2015 was not only the warmest year on record, it broke the record by the largest margin by which the record has been broken. 2015 was the 39th consecutive year with above-average temperatures. Ocean oscillations like El Niño Southern Oscillation (ENSO) can affect global average temperatures, for example, 1998 temperatures were significantly enhanced by strong El Niño conditions. 1998 remained the warmest year until 2005 and 2010 and the temperature of both of these years was enhanced by El Niño periods. The large margin by which 2015 is the warmest year is also attributed to another strong El Niño. However, 2014 was ENSO neutral. According to NOAA and NASA, 2015 had the warmest respective months on record for 10 out of the 12 months. The average temperature around the globe was 1.62 °F (0.90 °C) or 20% above the twentieth century average. In a first, December 2015 was also the first month to ever reach a temperature 2 degrees Fahrenheit above normal for the planet.

Initial Causes of Temperature Changes (External Forcings)

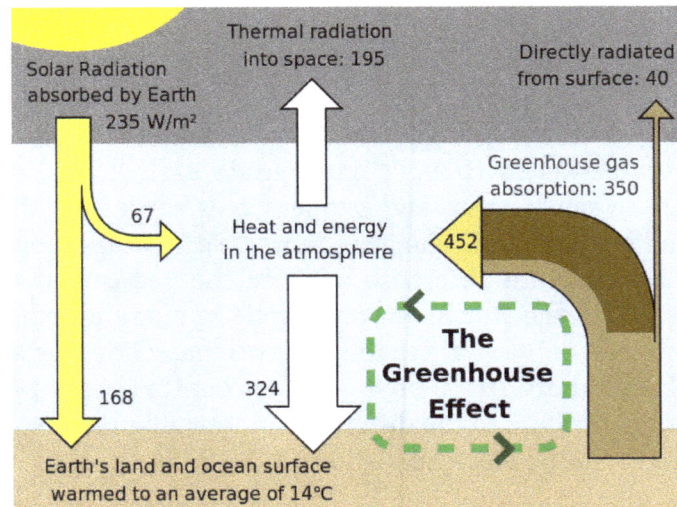

Greenhouse effect schematic showing energy flows between space, the atmosphere, and Earth's surface. Energy exchanges are expressed in watts per square meter (W/m²).

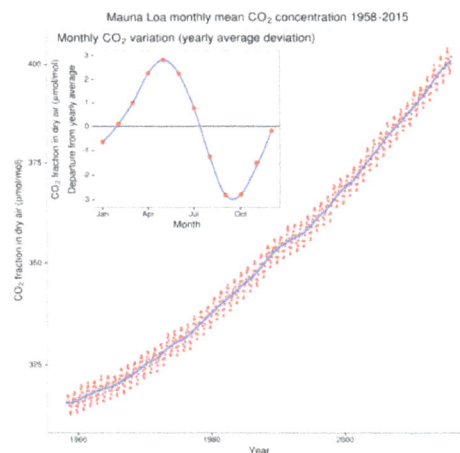

This graph, known as the Keeling Curve, documents the increase of atmospheric carbon dioxide concentrations from 1958–2015. Monthly CO_2 measurements display seasonal oscillations in an upward trend; each year's maximum occurs during the Northern Hemisphere's late spring, and declines during its growing season as plants remove some atmospheric CO_2.

The climate system can warm or cool in response to changes in *external forcings*. These are "external" to the climate system but not necessarily external to Earth. Examples of external forcings include changes in atmospheric composition (e.g., increased concentrations of greenhouse gases), solar luminosity, volcanic eruptions, and variations in Earth's orbit around the Sun.

Greenhouse Gases

The greenhouse effect is the process by which absorption and emission of infrared radiation by gases in a planet's atmosphere warm its lower atmosphere and surface. It was proposed by Joseph Fourier in 1824, discovered in 1860 by John Tyndall, was first investigated quantitatively by Svante Arrhenius in 1896, and was developed in the 1930s through 1960s by Guy Stewart Callendar.

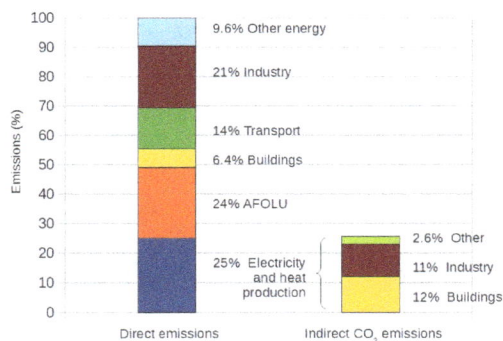

Annual world greenhouse gas emissions, in 2010, by sector.

Percentage share of global cumulative energy-related CO_2 emissions between 1751 and 2012 across different regions.

On Earth, naturally occurring amounts of greenhouse gases cause air temperature near the surface to be about 33 °C (59 °F) warmer than it would be in their absence.[d] Without the Earth's atmosphere, the Earth's average temperature would be well below the freezing temperature of water. The major greenhouse gases are water vapor, which causes about 36–70% of the greenhouse effect; carbon dioxide (CO_2), which causes 9–26%; methane (CH_4), which causes 4–9%; and ozone (O_3), which causes 3–7%. Clouds also affect the radiation balance through cloud forcings similar to greenhouse gases.

Human activity since the Industrial Revolution has increased the amount of greenhouse gases in the atmosphere, leading to increased radiative forcing from CO_2, methane, tropospheric ozone, CFCs and nitrous oxide. According to work published in 2007, the concentrations of CO_2 and methane have increased by 36% and 148% respectively since 1750. These levels are much higher than at any time during the last 800,000 years, the period for which reliable data has been extracted from ice cores. Less direct geological evidence indicates that CO_2 values higher than this were last seen about 20 million years ago.

Fossil fuel burning has produced about three-quarters of the increase in CO_2 from human activity over the past 20 years. The rest of this increase is caused mostly by changes in land-use, particularly deforestation. Another significant non-fuel source of anthropogenic CO_2 emissions is the calcination of limestone for clinker production, a chemical process which releases CO_2. Estimates of global CO_2 emissions in 2011 from fossil fuel combustion, including cement production and gas flaring, was 34.8 billion tonnes (9.5 ± 0.5 PgC), an increase of 54% above emissions in 1990. Coal burning was responsible for 43% of the total emissions, oil 34%, gas 18%, cement 4.9% and gas flaring 0.7%

In May 2013, it was reported that readings for CO_2 taken at the world's primary benchmark site in Mauna Loa surpassed 400 ppm. According to professor Brian Hoskins, this is likely the first time CO_2 levels have been this high for about 4.5 million years. Monthly global CO_2 concentrations exceeded 400 ppm in March 2015, probably for the first time in several million years. On 12 November 2015, NASA scientists reported that human-made carbon dioxide continues to increase above levels not seen in hundreds of thousands of years: currently, about half of the carbon dioxide released from the burning of fossil fuels is not absorbed by vegetation and the oceans and remains in the atmosphere.

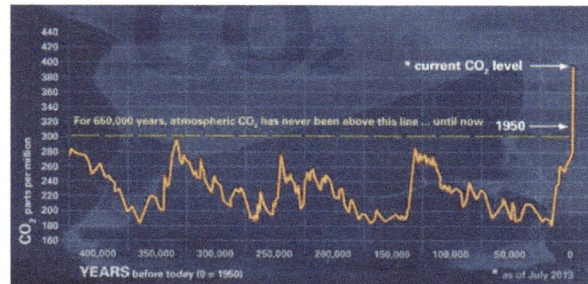

Atmospheric CO_2 concentration from 650,000 years ago to near present,
using ice core proxy data and direct measurements.

Over the last three decades of the twentieth century, gross domestic product per capita and population growth were the main drivers of increases in greenhouse gas emissions. CO_2 emissions are continuing to rise due to the burning of fossil fuels and land-use change. Emissions can be attributed to different regions. Attributions of emissions due to land-use change are subject to considerable uncertainty.

Emissions scenarios, estimates of changes in future emission levels of greenhouse gases, have been projected that depend upon uncertain economic, sociological, technological, and natural developments. In most scenarios, emissions continue to rise over the century, while in a few, emissions are reduced. Fossil fuel reserves are abundant, and will not limit carbon emissions in the 21st century. Emission scenarios, combined with modelling of the carbon cycle, have been used to produce estimates of how atmospheric concentrations of greenhouse gases might change in the future. Using the six IPCC SRES "marker" scenarios, models suggest that by the year 2100, the atmospheric concentration of CO_2 could range between 541 and 970 ppm. This is 90–250% above the concentration in the year 1750.

The popular media and the public often confuse global warming with ozone depletion, i.e., the destruction of stratospheric ozone (e.g., the ozone layer) by chlorofluorocarbons. Although there are a few areas of linkage, the relationship between the two is not strong. Reduced stratospheric ozone has had a slight cooling influence on surface temperatures, while increased tropospheric ozone has had a somewhat larger warming effect.

Aerosols and Soot

Global dimming, a gradual reduction in the amount of global direct irradiance at the Earth's surface, was observed from 1961 until at least 1990. Solid and liquid particles known as *aerosols*, produced by volcanoes and human-made pollutants, are thought to be the main cause of this dimming. They exert a cooling effect by increasing the reflection of incoming sunlight. The effects of the products of fossil fuel combustion – CO_2 and aerosols – have partially offset one another in recent decades, so that net warming has been due to the increase in non-CO_2 greenhouse gases such as methane. Radiative forcing due to aerosols is temporally limited due to the processes that remove aerosols from the atmosphere. Removal by clouds and precipitation gives tropospheric aerosols an atmospheric lifetime of only about a week, while stratospheric aerosols can remain for a few years. Carbon dioxide has a lifetime of a century or more, and as such, changes in aerosols will only delay climate changes due to carbon dioxide. Black carbon is second only to carbon dioxide for its contribution to global warming.

In addition to their direct effect by scattering and absorbing solar radiation, aerosols have indirect effects on the Earth's radiation budget. Sulfate aerosols act as cloud condensation nuclei and thus lead to clouds that have more and smaller cloud droplets. These clouds reflect solar radiation more efficiently than clouds with fewer and larger droplets, a phenomenon known as the Twomey effect. This effect also causes droplets to be of more uniform size, which reduces growth of raindrops and makes the cloud more reflective to incoming sunlight, known as the Albrecht effect. Indirect effects are most noticeable in marine stratiform clouds, and have very little radiative effect on convective clouds. Indirect effects of aerosols represent the largest uncertainty in radiative forcing.

Soot may either cool or warm Earth's climate system, depending on whether it is airborne or deposited. Atmospheric soot directly absorbs solar radiation, which heats the atmosphere and cools the surface. In isolated areas with high soot production, such as rural India, as much as 50% of surface warming due to greenhouse gases may be masked by atmospheric brown clouds. When deposited, especially on glaciers or on ice in arctic regions, the lower surface albedo can also directly heat the surface. The influences of atmospheric particles, including black carbon, are most pronounced in the tropics and sub-tropics, particularly in Asia, while the effects of greenhouse gases are dominant in the extratropics and southern hemisphere.

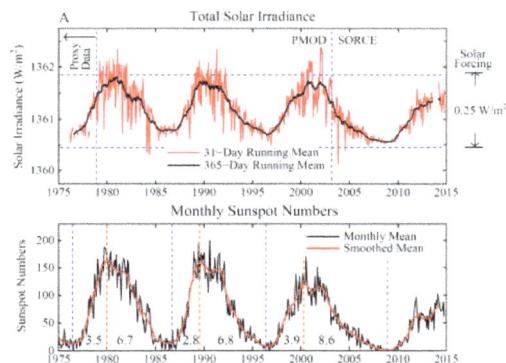

Changes in Total Solar Irradiance (TSI) and monthly sunspot numbers since the mid-1970s.

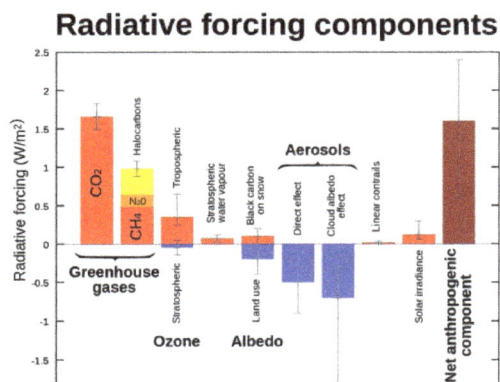

Contribution of natural factors and human activities to radiative forcing of climate change. Radiative forcing values are for the year 2005, relative to the pre-industrial era (1750). The contribution of solar irradiance to radiative forcing is 5% the value of the combined radiative forcing due to increases in the atmospheric concentrations of carbon dioxide, methane and nitrous oxide.

Solar Activity

Since 1978, solar irradiance has been measured by satellites. These measurements indicate that the Sun's radiative output has not increased since 1978, so the warming during the past 30 years cannot be attributed to an increase in solar energy reaching the Earth.

Climate models have been used to examine the role of the Sun in recent climate change. Models are unable to reproduce the rapid warming observed in recent decades when they only take into account variations in solar output and volcanic activity. Models are, however, able to simulate the observed 20th century changes in temperature when they include all of the most important external forcings, including human influences and natural forcings.

Another line of evidence against solar variations having caused recent climate change comes from looking at how temperatures at different levels in the Earth's atmosphere have changed. Models and observations show that greenhouse warming results in warming of the lower atmosphere (the troposphere) but cooling of the upper atmosphere (the stratosphere). Depletion of the ozone layer by chemical refrigerants has also resulted in a strong cooling effect in the stratosphere. If solar variations were responsible for observed warming, warming of both the troposphere and stratosphere would be expected.

Variations in Earth's Orbit

The tilt of the Earth's axis and the shape of its orbit around the Sun vary slowly over tens of thousands of years and are a natural source of climate change, by changing the seasonal and latitudinal distribution of solar insolation.

During the last few thousand years, this phenomenon contributed to a slow cooling trend at high latitudes of the Northern Hemisphere during summer, a trend that was reversed by greenhouse-gas-induced warming during the 20th century.

Variations in orbital cycles may initiate a new glacial period in the future, though the timing of this depends on greenhouse gas concentrations as well as the orbital forcing. A new glacial period is not expected within the next 50,000 years if atmospheric CO_2 concentration remains above 300 ppm.

Feedback

The climate system includes a range of *feedbacks*, which alter the response of the system to changes in external forcings. Positive feedbacks increase the response of the climate system to an initial forcing, while negative feedbacks reduce it.

There are a range of feedbacks in the climate system, including water vapor, changes in ice-albedo (snow and ice cover affect how much the Earth's surface absorbs or reflects incoming sunlight), clouds, and changes in the Earth's carbon cycle (e.g., the release of carbon from soil). The main negative feedback is the energy the Earth's surface radiates into space as infrared radiation. According to the Stefan-Boltzmann law, if the absolute temperature (as measured in kelvin) doubles,[e] radiated energy increases by a factor of 16 (2 to the 4th power).

Sea ice, shown here in Nunavut, in northern Canada, reflects more sunshine, while open ocean absorbs more, accelerating melting.

Feedbacks are an important factor in determining the sensitivity of the climate system to increased atmospheric greenhouse gas concentrations. Other factors being equal, a higher *climate sensitivity* means that more warming will occur for a given increase in greenhouse gas forcing. Uncertainty over the effect of feedbacks is a major reason why different climate models project different magnitudes of warming for a given forcing scenario. More research is needed to understand the role of clouds and carbon cycle feedbacks in climate projections.

The IPCC projections previously mentioned span the "likely" range (greater than 66% probability, based on expert judgement) for the selected emissions scenarios. However, the IPCC's projections do not reflect the full range of uncertainty. The lower end of the "likely" range appears to be better constrained than the upper end.

Climate Models

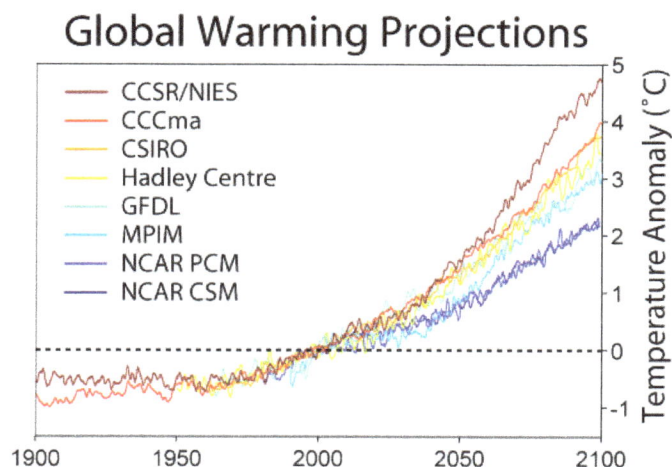

Calculations of global warming prepared in or before 2001 from a range of climate models under the SRES A2 emissions scenario, which assumes no action is taken to reduce emissions and regionally divided economic development.

Projected change in annual mean surface air temperature from the late 20th century to the middle 21st century, based on a medium emissions scenario (SRES A1B). This scenario assumes that no future policies are adopted to limit greenhouse gas emissions. Image credit: NOAA GFDL.

A climate model is a representation of the physical, chemical and biological processes that affect the climate system. Such models are based on scientific disciplines such as fluid dynamics and thermodynamics as well as physical processes such as radiative transfer. The models may be used to predict a range of variables such as local air movement, temperature, clouds, and other atmospheric properties; ocean temperature, salt content, and circulation; ice cover on land and sea; the transfer of heat and moisture from soil and vegetation to the atmosphere; and chemical and biological processes, among others.

Although researchers attempt to include as many processes as possible, simplifications of the actual climate system are inevitable because of the constraints of available computer power and limitations in knowledge of the climate system. Results from models can also vary due to different greenhouse gas inputs and the model's climate sensitivity. For example, the uncertainty in IPCC's 2007 projections is caused by (1) the use of multiple models with differing sensitivity to greenhouse gas concentrations, (2) the use of differing estimates of humanity's future greenhouse gas emissions, (3) any additional emissions from climate feedbacks that were not included in the models IPCC used to prepare its report, i.e., greenhouse gas releases from permafrost.

The models do not assume the climate will warm due to increasing levels of greenhouse gases. Instead the models predict how greenhouse gases will interact with radiative transfer and other physical processes. Warming or cooling is thus a result, not an assumption, of the models.

Clouds and their effects are especially difficult to predict. Improving the models' representation of clouds is therefore an important topic in current research. Another prominent research topic is expanding and improving representations of the carbon cycle.

Models are also used to help investigate the causes of recent climate change by comparing the observed changes to those that the models project from various natural and human causes. Although these models do not unambiguously attribute the warming that occurred from approximately 1910 to 1945 to either natural variation or human effects, they do indicate that the warming since 1970 is dominated by anthropogenic greenhouse gas emissions.

The physical realism of models is tested by examining their ability to simulate contemporary or past climates. Climate models produce a good match to observations of global temperature changes over the last century, but do not simulate all aspects of climate. Not all effects of global warming are accurately predicted by the climate models used by the IPCC. Observed Arctic shrinkage has been faster than that predicted. Precipitation increased proportionally to atmospheric humidity, and hence significantly faster than global climate models predict. Since 1990, sea level has also risen considerably faster than models predicted it would.

Observed and Expected Environmental Effects

Anthropogenic forcing has likely contributed to some of the observed changes, including sea level rise, changes in climate extremes (such as the number of warm and cold days), declines in Arctic sea ice extent, glacier retreat, and greening of the Sahara.

During the 21st century, glaciers and snow cover are projected to continue their widespread retreat. Projections of declines in Arctic sea ice vary. Recent projections suggest that Arctic summers could be ice-free (defined as ice extent less than 1 million square km) as early as 2025-2030.

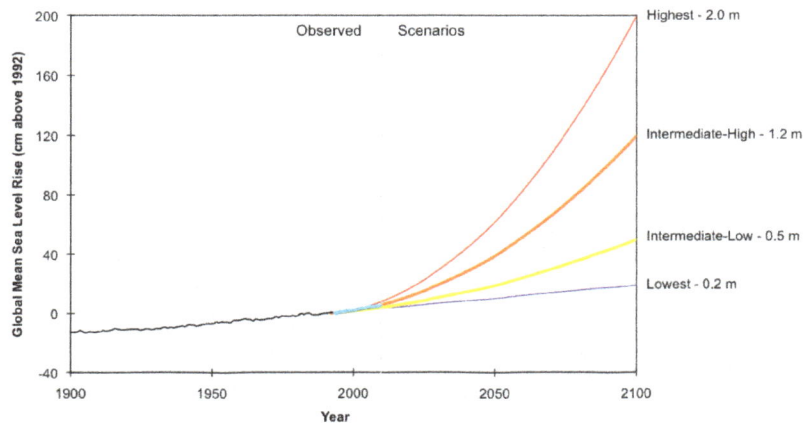

Projections of global mean sea level rise by Parris and others. Probabilities have not been assigned to these projections. Therefore, none of these projections should be interpreted as a "best estimate" of future sea level rise. Image credit: NOAA.

"Detection" is the process of demonstrating that climate has changed in some defined statistical sense, without providing a reason for that change. Detection does not imply attribution of the detected change to a particular cause. "Attribution" of causes of climate change is the process of establishing the most likely causes for the detected change with some defined level of confidence. Detection and attribution may also be applied to observed changes in physical, ecological and social systems.

Extreme Weather

Changes in regional climate are expected to include greater warming over land, with most warming at high northern latitudes, and least warming over the Southern Ocean and parts of the North Atlantic Ocean.

Future changes in precipitation are expected to follow existing trends, with reduced precipitation over subtropical land areas, and increased precipitation at subpolar latitudes and some equatorial regions. Projections suggest a probable increase in the frequency and severity of some extreme weather events, such as heat waves.

A 2015 study published in *Nature Climate Change*, states:

> About 18% of the moderate daily precipitation extremes over land are attributable to the observed temperature increase since pre-industrial times, which in turn primarily results from human influence. For 2 °C of warming the fraction of precipitation extremes attributable to human influence rises to about 40%. Likewise, today about 75% of the moderate daily hot extremes over land are attributable to warming. It is the most rare and extreme events for which the largest fraction is anthropogenic, and that contribution increases nonlinearly with further warming.

Data analysis of extreme events from 1960 till 2010 suggests that droughts and heat waves appear simultaneously with increased frequency. Extremely wet or dry events within the monsoon period have increased since 1980.

Sea Level Rise

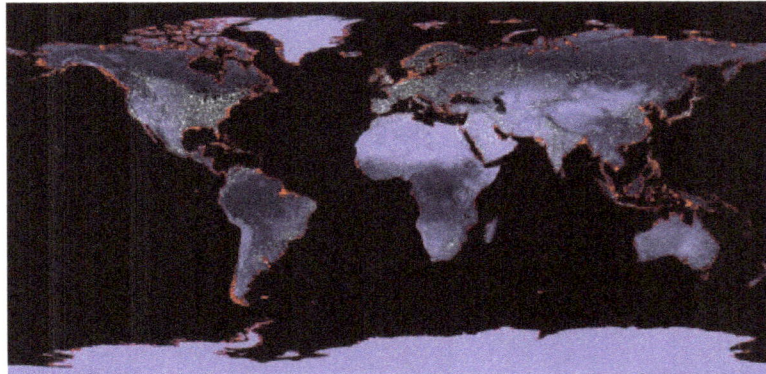

Map of the Earth with a six-meter sea level rise represented in red.

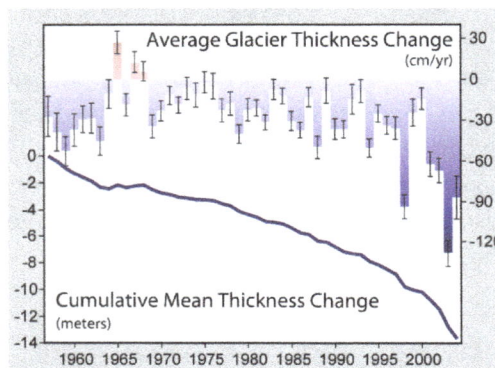

Sparse records indicate that glaciers have been retreating since the early 1800s. In the 1950s measurements began that allow the monitoring of glacial mass balance, reported to the World Glacier Monitoring Service (WGMS) and the National Snow and Ice Data Center (NSIDC).

The sea level rise since 1993 has been estimated to have been on average 2.6 mm and 2.9 mm per year ± 0.4 mm. Additionally, sea level rise has accelerated from 1995 to 2015. Over the 21st century, the IPCC projects for a high emissions scenario, that global mean sea level could rise by 52–98 cm. The IPCC's projections are conservative, and may underestimate future sea level rise. Other estimates suggest that for the same period, global mean sea level could rise by 0.2 to 2.0 m (0.7–6.6 ft), relative to mean sea level in 1992.

Widespread coastal flooding would be expected if several degrees of warming is sustained for millennia. For example, sustained global warming of more than 2 °C (relative to pre-industrial levels) could lead to eventual sea level rise of around 1 to 4 m due to thermal expansion of sea water and the melting of glaciers and small ice caps. Melting of the Greenland ice sheet could contribute an additional 4 to 7.5 m over many thousands of years. It has been estimated that we are already committed to a sea-level rise of approximately 2.3 meters for each degree of temperature rise within the next 2,000 years.

Warming beyond the 2 °C target would potentially lead to rates of sea-level rise dominated by ice loss from Antarctica. Continued CO_2 emissions from fossil sources could cause additional tens of meters of sea level rise, over the next millennia and eventually ultimately eliminate the entire Antarctic ice sheet, causing about 58 meters of sea level rise.

Ecological Systems

In terrestrial ecosystems, the earlier timing of spring events, as well as poleward and upward shifts in plant and animal ranges, have been linked with high confidence to recent warming. Future climate change is expected to affect particular ecosystems, including tundra, mangroves, and coral reefs. It is expected that most ecosystems will be affected by higher atmospheric CO_2 levels, combined with higher global temperatures. Overall, it is expected that climate change will result in the extinction of many species and reduced diversity of ecosystems.

Increases in atmospheric CO_2 concentrations have led to an increase in ocean acidity. Dissolved CO_2 increases ocean acidity, measured by lower pH values. Between 1750 and 2000, surface-ocean pH has decreased by ≈0.1, from ≈8.2 to ≈8.1. Surface-ocean pH has probably not been below ≈8.1 during the past 2 million years. Projections suggest that surface-ocean pH could decrease by an additional 0.3–0.4 units by 2100. Future ocean acidification could threaten coral reefs, fisheries, protected species, and other natural resources of value to society.

Ocean deoxygenation is projected to increase hypoxia by 10%, and triple suboxic waters (oxygen concentrations 98% less than the mean surface concentrations), for each 1 °C of upper ocean warming.

Long-term Effects

On the timescale of centuries to millennia, the magnitude of global warming will be determined primarily by anthropogenic CO_2 emissions. This is due to carbon dioxide's very long lifetime in the atmosphere.

Stabilizing the global average temperature would require large reductions in CO_2 emissions, as well as reductions in emissions of other greenhouse gases such as methane and nitrous oxide. Emissions of CO_2 would need to be reduced by more than 80% relative to their peak level. Even if this were achieved, global average temperatures would remain close to their highest level for many centuries.

Long-term effects also include a response from the Earth's crust, due to ice melting and deglaciation, in a process called post-glacial rebound, when land masses are no longer depressed by the weight of ice. This could lead to landslides and increased seismic and volcanic activities. Tsunamis could be generated by submarine landslides caused by warmer ocean water thawing ocean-floor permafrost or releasing gas hydrates. Some world regions, such as the French Alps, already show signs of an increase in landslide frequency.

Large-Scale and Abrupt Impacts

Climate change could result in global, large-scale changes in natural and social systems. Examples include the possibility for the Atlantic Meridional Overturning Circulation to slow- or shutdown, which in the instance of a shutdown would change weather in Europe and North America considerably, ocean acidification caused by increased atmospheric concentrations of carbon dioxide, and the long-term melting of ice sheets, which contributes to sea level rise.

Some large-scale changes could occur abruptly, i.e., over a short time period, and might also be irreversible. Examples of abrupt climate change are the rapid release of methane and carbon dioxide from permafrost, which would lead to amplified global warming, or the shutdown of thermohaline circulation. Scientific understanding of abrupt climate change is generally poor. The probability of abrupt change for some climate related feedbacks may be low. Factors that may increase the probability of abrupt climate change include higher magnitudes of global warming, warming that occurs more rapidly, and warming that is sustained over longer time periods.

Observed and Expected Effects on Social Systems

The effects of climate change on human systems, mostly due to warming or shifts in precipitation patterns, or both, have been detected worldwide. Production of wheat and maize globally has been impacted by climate change. While crop production has increased in some mid-latitude regions such as the UK and Northeast China, economic losses due to extreme weather events have increased globally. There has been a shift from cold- to heat-related mortality in some regions as a result of warming. Livelihoods of indigenous peoples of the Arctic have been altered by climate change, and there is emerging evidence of climate change impacts on livelihoods of indigenous peoples in other regions. Regional impacts of climate change are now observable at more locations than before, on all continents and across ocean regions.

The future social impacts of climate change will be uneven. Many risks are expected to increase with higher magnitudes of global warming. All regions are at risk of experiencing negative impacts. Low-latitude, less developed areas face the greatest risk. A study from 2015 concluded that economic growth (gross domestic product) of poorer countries is much more impaired with projected future climate warming, than previously thought.

A meta-analysis of 56 studies concluded in 2014 that each degree of temperature rise will increase violence by up to 20%, which includes fist fights, violent crimes, civil unrest or wars.

Examples of impacts include:

- *Food*: Crop production will probably be negatively affected in low latitude countries, while effects at northern latitudes may be positive or negative. Global warming of around 4.6 °C relative to pre-industrial levels could pose a large risk to global and regional food security.

- *Health*: Generally impacts will be more negative than positive. Impacts include: the effects of extreme weather, leading to injury and loss of life; and indirect effects, such as undernutrition brought on by crop failures.

Habitat Inundation

In small islands and mega deltas, inundation as a result of sea level rise is expected to threaten vital infrastructure and human settlements. This could lead to issues of homelessness in countries with low-lying areas such as Bangladesh, as well as statelessness for populations in countries such as the Maldives and Tuvalu.

Economy

Estimates based on the IPCC A1B emission scenario from additional CO_2 and CH_4 greenhouse gases released from permafrost, estimate associated impact damages by US$43 trillion.

Infrastructure

Continued permafrost degradation will likely result in unstable infrastructure in Arctic regions, or Alaska before 2100. Thus, impacting roads, pipelines and buildings, as well as water distribution, and cause slope failures.

Possible Responses to Global Warming

Mitigation

The graph on the right shows three "pathways" to meet the UNFCCC's 2 °C target, labelled "global technology", "decentralised solutions", and "consumption change". Each pathway shows how various measures (e.g., improved energy efficiency, increased use of renewable energy) could contribute to emissions reductions. Image credit: PBL Netherlands Environmental Assessment Agency.

Mitigation of climate change are actions to reduce greenhouse gas emissions, or enhance the capacity of carbon sinks to absorb GHGs from the atmosphere. There is a large potential for future reductions in emissions by a combination of activities, including: energy conservation and increased energy efficiency; the use of low-carbon energy technologies, such as renewable energy, nuclear energy, and carbon capture and storage; and enhancing carbon sinks through, for example, reforestation and preventing deforestation. A 2015 report by Citibank concluded that transitioning to a low carbon economy would yield positive return on investments.

Near- and long-term trends in the global energy system are inconsistent with limiting global warming at below 1.5 or 2 °C, relative to pre-industrial levels. Pledges made as part of the Cancún agreements are broadly consistent with having a likely chance (66 to 100% probability) of limiting global warming (in the 21st century) at below 3 °C, relative to pre-industrial levels.

In limiting warming at below 2 °C, more stringent emission reductions in the near-term would allow for less rapid reductions after 2030. Many integrated models are unable to meet the 2 °C target if pessimistic assumptions are made about the availability of mitigation technologies.

Adaptation

Other policy responses include adaptation to climate change. Adaptation to climate change may be planned, either in reaction to or anticipation of climate change, or spontaneous, i.e., without government intervention. Planned adaptation is already occurring on a limited basis. The barriers, limits, and costs of future adaptation are not fully understood.

A concept related to adaptation is *adaptive capacity*, which is the ability of a system (human, natural or managed) to adjust to climate change (including climate variability and extremes) to moderate potential damages, to take advantage of opportunities, or to cope with consequences. Unmitigated climate change (i.e., future climate change without efforts to limit greenhouse gas emissions) would, in the long term, be likely to exceed the capacity of natural, managed and human systems to adapt.

Environmental organizations and public figures have emphasized changes in the climate and the risks they entail, while promoting adaptation to changes in infrastructural needs and emissions reductions.

Climate Engineering

Climate engineering (sometimes called *geoengineering* or *climate intervention*) is the deliberate modification of the climate. It has been investigated as a possible response to global warming, e.g. by NASA and the Royal Society. Techniques under research fall generally into the categories solar radiation management and carbon dioxide removal, although various other schemes have been suggested. A study from 2014 investigated the most common climate engineering methods and concluded they are either ineffective or have potentially severe side effects and cannot be stopped without causing rapid climate change.

Discourse about Global Warming

Political Discussion

Further information: 2011 United Nations Climate Change Conference, 2012 United Nations Climate Change Conference, 2013 United Nations Climate Change Conference, and 2015 United Nations Climate Change Conference

Article 2 of the UN Framework Convention refers explicitly to "stabilization of greenhouse gas concentrations." To stabilize the atmospheric concentration of CO_2, emissions worldwide would need to be dramatically reduced from their present level.

Most countries in the world are parties to the United Nations Framework Convention on Climate Change (UNFCCC). The ultimate objective of the Convention is to prevent dangerous human interference of the climate system. As stated in the Convention, this requires that GHG concentrations are stabilized in the atmosphere at a level where ecosystems can adapt naturally to climate change, food production is not threatened, and economic development can proceed in a sustainable fashion. The Framework Convention was agreed in 1992, but since then, global emissions have risen.

During negotiations, the G77 (a lobbying group in the United Nations representing 133 developing nations) pushed for a mandate requiring developed countries to "[take] the lead" in reducing their emissions. This was justified on the basis that: the developed world's emissions had contributed most to the cumulation of GHGs in the atmosphere; per-capita emissions (i.e., emissions per head of population) were still relatively low in developing countries; and the emissions of developing countries would grow to meet their development needs.

This mandate was sustained in the Kyoto Protocol to the Framework Convention, which entered into legal effect in 2005. In ratifying the Kyoto Protocol, most developed countries accepted legally binding commitments to limit their emissions. These first-round commitments expired in 2012. United States President George W. Bush rejected the treaty on the basis that "it exempts 80% of the world, including major population centers such as China and India, from compliance, and would cause serious harm to the US economy."

At the 15th UNFCCC Conference of the Parties, held in 2009 at Copenhagen, several UNFCCC Parties produced the Copenhagen Accord. Parties associated with the Accord (140 countries, as of November 2010) aim to limit the future increase in global mean temperature to below 2°C. The 16th Conference of the Parties (COP16) was held at Cancún in 2010. It produced an agreement, not a binding treaty, that the Parties should take urgent action to reduce greenhouse gas emissions to meet a goal of limiting global warming to 2°C above pre-industrial temperatures. It also recognized the need to consider strengthening the goal to a global average rise of 1.5 °C.

Scientific Discussion

There is continuing discussion through published peer-reviewed scientific papers, which are assessed by scientists working in the relevant fields taking part in the Intergovernmental Panel on Climate Change. The scientific consensus as of 2013 stated in the IPCC Fifth Assessment Report is that it "is extremely likely that human influence has been the dominant cause of the observed warming since the mid-20th century". A 2008 report by the U.S. National Academy of Sciences stated that most scientists by then agreed that observed warming in recent decades was primarily caused by human activities increasing the amount of greenhouse gases in the atmosphere. In 2005 the Royal Society stated that while the overwhelming majority of scientists were in agreement on the main points, some individuals and organisations opposed to the consensus on urgent action needed to reduce greenhouse gas emissions have tried to undermine the science and work of the IPCC. National science academies have called on world leaders for policies to cut global emissions.

In the scientific literature, there is a strong consensus that global surface temperatures have increased in recent decades and that the trend is caused mainly by human-induced emissions of greenhouse gases. No scientific body of national or international standing disagrees with this view.

Discussion by the Public and in Popular Media

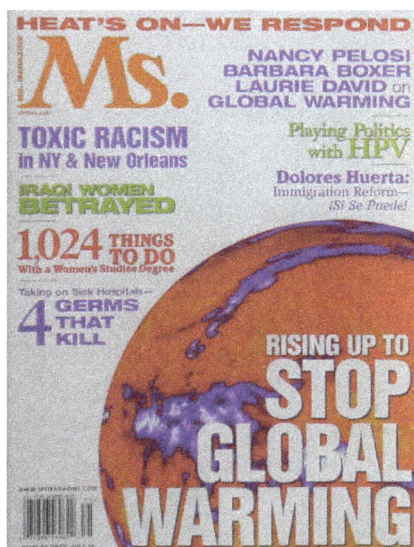

Global warming was the cover story in this 2007 issue of *Ms. magazine*

The global warming controversy refers to a variety of disputes, substantially more pronounced in the popular media than in the scientific literature, regarding the nature, causes, and consequences of global warming. The disputed issues include the causes of increased global average air temperature, especially since the mid-20th century, whether this warming trend is unprecedented or within normal climatic variations, whether humankind has contributed significantly to it, and whether the increase is completely or partially an artifact of poor measurements. Additional disputes concern estimates of climate sensitivity, predictions of additional warming, and what the consequences of global warming will be.

From 1990 to 1997, right-wing conservative think tanks in the United States mobilized to challenge the legitimacy of global warming as a social problem. They challenged the scientific evidence, argued that global warming will have benefits, and asserted that proposed solutions would do more harm than good. Some people dispute aspects of climate change science. Organizations such as the libertarian Competitive Enterprise Institute, conservative commentators, and some companies such as ExxonMobil have challenged IPCC climate change scenarios, funded scientists who disagree with the scientific consensus, and provided their own projections of the economic cost of stricter controls. On the other hand, some fossil fuel companies have scaled back their efforts in recent years, or even called for policies to reduce global warming. Global oil companies have begun to acknowledge climate change exists and is caused by human activities and the burning of fossil fuels.

Surveys of Public Opinion

The world public, or at least people in economically advanced regions, became broadly aware of the global warming problem in the late 1980s. Polling groups began to track opinions on the subject, at first mainly in the United States. The longest consistent polling, by Gallup in the US, found relatively small deviations of 10% or so from 1998 to 2015 in opinion on the seriousness of global warming, but with increasing polarization between those concerned and those unconcerned.

The first major worldwide poll, conducted by Gallup in 2008-2009 in 127 countries, found that some 62% of people worldwide said they knew about global warming. In the advanced countries of North America, Europe and Japan, 90% or more knew about it (97% in the U.S., 99% in Japan); in less developed countries, especially in Africa, fewer than a quarter knew about it, although many had noticed local weather changes. Among those who knew about global warming, there was a wide variation between nations in belief that the warming was a result of human activities.

By 2010, with 111 countries surveyed, Gallup determined that there was a substantial decrease since 2007–08 in the number of Americans and Europeans who viewed global warming as a serious threat. In the US, just a little over half the population (53%) now viewed it as a serious concern for either themselves or their families; this was 10 points below the 2008 poll (63%). Latin America had the biggest rise in concern: 73% said global warming is a serious threat to their families. This global poll also found that people are more likely to attribute global warming to human activities than to natural causes, except in the US where nearly half (47%) of the population attributed global warming to natural causes.

A March–May 2013 survey by Pew Research Center for the People & the Press polled 39 countries about global threats. According to 54% of those questioned, global warming featured top of the perceived global threats. In a January 2013 survey, Pew found that 69% of Americans say there is solid evidence that the Earth's average temperature has got warmer over the past few decades, up six points since November 2011 and 12 points since 2009.

A 2010 survey of 14 industrialized countries found that skepticism about the danger of global warming was highest in Australia, Norway, New Zealand and the United States, in that order, correlating positively with per capita emissions of carbon dioxide.

Etymology

In the 1950s, research suggested increasing temperatures, and a 1952 newspaper reported "climate change". This phrase next appeared in a November 1957 report in *The Hammond Times* which described Roger Revelle's research into the effects of increasing human-caused CO_2 emissions on the greenhouse effect, "a large scale global warming, with radical climate changes may result". Both phrases were only used occasionally until 1975, when Wallace Smith Broecker published a scientific paper on the topic; "Climatic Change: Are We on the Brink of a Pronounced Global Warming?" The phrase began to come into common use, and in 1976 Mikhail Budyko's statement that "a global warming up has started" was widely reported. Other studies, such as a 1971 MIT report, referred to the human impact as "inadvertent climate modification", but an influential 1979 National Academy of Sciences study headed by Jule Charney followed Broecker in using *global warming* for rising surface temperatures, while describing the wider effects of increased CO_2 as *climate change*.

In 1986 and November 1987, NASA climate scientist James Hansen gave testimony to Congress on global warming. There were increasing heatwaves and drought problems in the summer of 1988, and when Hansen testified in the Senate on 23 June he sparked worldwide interest. He said: "global warming has reached a level such that we can ascribe with a high degree of confidence a cause and effect relationship between the greenhouse effect and the observed warming." Public attention increased over the summer, and *global warming* became the dominant popular term, commonly used both by the press and in public discourse.

In a 2008 NASA article on usage, Erik M. Conway defined *Global warming* as "the increase in Earth's average surface temperature due to rising levels of greenhouse gases", while *Climate change* was "a long-term change in the Earth's climate, or of a region on Earth." As effects such as changing patterns of rainfall and rising sea levels would probably have more impact than temperatures alone, he considered *global climate change* a more scientifically accurate term, and like the Intergovernmental Panel on Climate Change, the NASA website would emphasise this wider context.

References

- IEA (2009). World Energy Outlook 2009 (PDF). Paris, France: International Energy Agency (IEA). ISBN 978-92-64-06130-9.

- IPCC AR4 SYR (2007). Core Writing Team; Pachauri, R.K; Reisinger, A., eds. Climate Change 2007: Synthesis Report. Contribution of Working Groups I, II and III to the Fourth Assessment Report of the Intergovernmental Panel on Climate Change. IPCC. ISBN 92-9169-122-4.

- IPCC AR4 WG3 (2007). Metz, B.; Davidson, O.R.; Bosch, P.R.; Dave, R.; Meyer, L.A., eds. Climate Change 2007: Mitigation of Climate Change. Contribution of Working Group III to the Fourth Assessment Report of the Intergovernmental Panel on Climate Change. Cambridge University Press. ISBN 978-0-521-88011-4.

- IPCC SAR WG3 (1996). Bruce, J.P.; Lee, H.; Haites, E.F., eds. Climate Change 1995: Economic and Social Dimensions of Climate Change. Contribution of Working Group III to the Second Assessment Report of the Intergovernmental Panel on Climate Change. Cambridge University Press. ISBN 0-521-56051-9.

- IPCC TAR WG1 (2001). Houghton, J.T.; Ding, Y.; Griggs, D.J.; Noguer, M.; van der Linden, P.J.; Dai, X.; Maskell, K.; Johnson, C.A., eds. Climate Change 2001: The Scientific Basis. Contribution of Working Group I to the Third Assessment Report of the Intergovernmental Panel on Climate Change. Cambridge University Press. ISBN 0-521-80767-0.

- IPCC TAR WG2 (2001). McCarthy, J. J.; Canziani, O. F.; Leary, N. A.; Dokken, D. J.; White, K. S., eds. Climate Change 2001: Impacts, Adaptation and Vulnerability. Contribution of Working Group II to the Third Assessment Report of the Intergovernmental Panel on Climate Change. Cambridge University Press. ISBN 0-521-80768-9.

- National Research Council (2010). America's Climate Choices: Panel on Advancing the Science of Climate Change;. Washington, D.C.: The National Academies Press. ISBN 0-309-14588-0.

- Parris, A.; et al. (6 December 2012), Global Sea Level Rise Scenarios for the US National Climate Assessment. NOAA Tech Memo OAR CPO-1 (PDF), NOAA Climate Program Office. Report website.

Causes of Global Warming

It can be proven outright that the exploitation of the Earth's natural resources by human activity has led to uncontrollable levels of greenhouse gas emissions. These emissions have to be systematically shut down in order to mitigate climate change. This chapter lists the major causes of global warming and how they are interconnected with each other.

Attribution of Recent Climate Change

Global annual average temperature; year-to-year fluctuations are due to natural processes, such as the effects of El Niños, La Niñas, and the eruption of large volcanoes.

Attribution of Surface Temperature trends since 1950

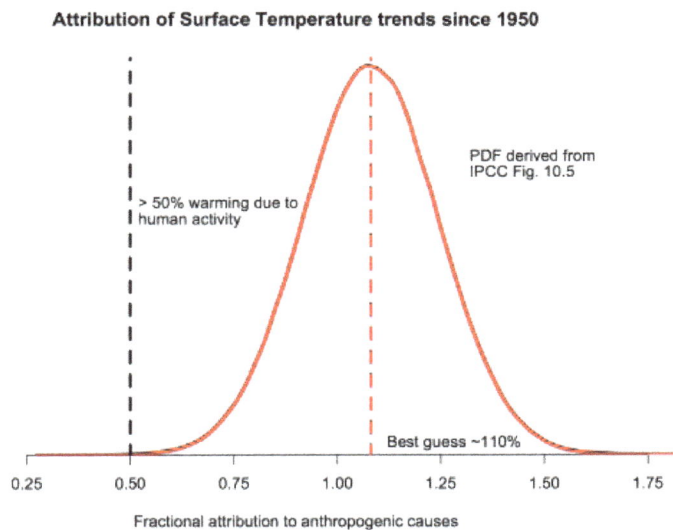

PDF derived from IPCC Fig. 10.5

> 50% warming due to human activity

Best guess ~110%

Fractional attribution to anthropogenic causes

PDF of fraction of surface temperature trends since 1950 attributable to human activity, based on IPCC AR5 10.5

Attribution of recent climate change is the effort to scientifically ascertain mechanisms responsible for recent climate changes on Earth, commonly known as 'global warming'. The effort has focused on changes observed during the period of instrumental temperature record, when records are most reliable; particularly in the last 50 years, when human activity has grown fastest and observations of the troposphere have become available. The dominant mechanisms (to which the IPCC attributes climate change) are anthropogenic, i.e., the result of human activity. They are:

- increasing atmospheric concentrations of greenhouse gases

- global changes to land surface, such as deforestation

- increasing atmospheric concentrations of aerosols.

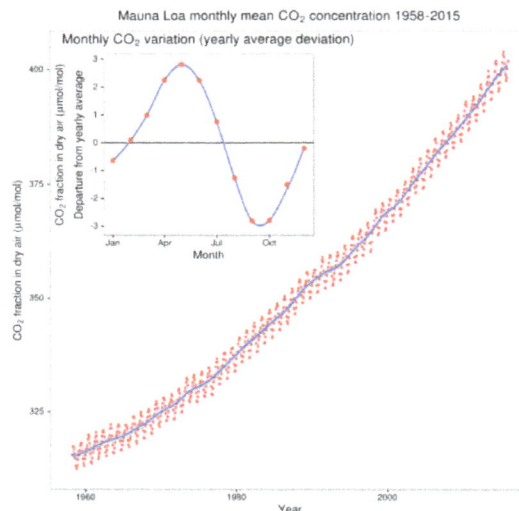

Mauna Loa monthly mean CO₂ concentration 1958-2015

This graph is known as the Keeling Curve and shows the long-term increase of atmospheric carbon dioxide (CO_2) concentrations from 1958–2015. Monthly CO_2 measurements display seasonal oscillations in an upward trend. Each year's maximum occurs during the Northern Hemisphere's late spring, and declines during its growing season as plants remove some atmospheric CO_2.
This image shows three examples of internal climate variability measured between 1950 and 2012: the El Niño–Southern oscillation, the Arctic oscillation, and the North Atlantic oscillation.

There are also natural mechanisms for variation including climate oscillations, changes in solar activity, and volcanic activity.

According to the Intergovernmental Panel on Climate Change (IPCC), it is "extremely likely" that human influence was the dominant cause of global warming between 1951 and 2010. The IPCC defines "extremely likely" as indicating a probability of 95 to 100%, based on an expert assessment of all the available evidence.

Multiple lines of evidence support attribution of recent climate change to human activities:

- A basic physical understanding of the climate system: greenhouse gas concentrations have increased and their warming properties are well-established.

- Historical estimates of past climate changes suggest that the recent changes in global surface temperature are unusual.

- Computer-based climate models are unable to replicate the observed warming unless human greenhouse gas emissions are included.

- Natural forces alone (such as solar and volcanic activity) cannot explain the observed warming.

The IPCC's attribution of recent global warming to human activities is a view shared by most scientists, and is also supported by 196 other scientific organizations worldwide.

Background

This section introduces some concepts in climate science that are used in the following sections:

Factors affecting Earth's climate can be broken down into feedbacks and forcings. A forcing is something that is imposed externally on the climate system. External forcings include natural phenomena such as volcanic eruptions and variations in the sun's output. Human activities can also impose forcings, for example, through changing the composition of the atmosphere.

Radiative forcing is a measure of how various factors alter the energy balance of the Earth's atmosphere. A positive radiative forcing will tend to increase the energy of the Earth-atmosphere system, leading to a warming of the system. Between the start of the Industrial Revolution in 1750, and the year 2005, the increase in the atmospheric concentration of carbon dioxide (chemical formula: CO_2) led to a positive radiative forcing, averaged over the Earth's surface area, of about 1.66 watts per square metre (abbreviated W m^{-2}).

Climate feedbacks can either amplify or dampen the response of the climate to a given forcing. There are many feedback mechanisms in the climate system that can either amplify (a positive feedback) or diminish (a negative feedback) the effects of a change in climate forcing.

Aspects of the climate system will show variation in response to changes in forcings. In the absence of forcings imposed on it, the climate system will still show internal variability. This internal variability is a result of complex interactions between components of the climate system, such as the coupling between the atmosphere and ocean. An example of internal variability is the El Niño-Southern Oscillation.

Detection vs. Attribution

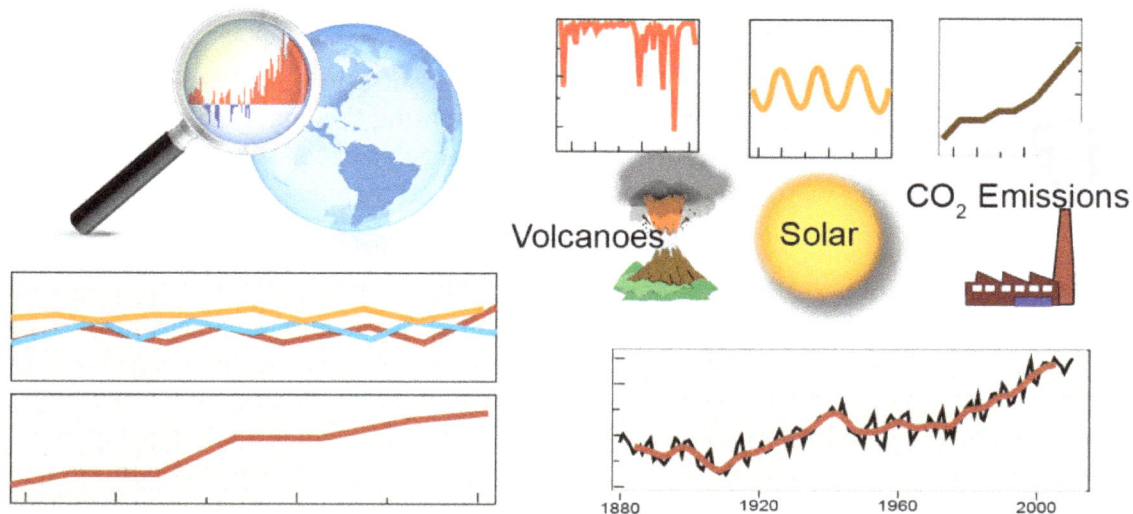

Detection and Attribution as Forensics

Volcanoes   Solar CO_2 Emissions

1880   1920   1960   2000

Detection: finding something out of the ordinary – a "signal" emerging from the noise

Attribution: determining the cause of the detected trend

In detection and attribution, the natural factors considered usually include changes in the Sun's output and volcanic eruptions, as well as natural modes of variability such as El Niño and La Niña. Human factors include the emissions of heat-trapping "greenhouse" gases and particulates as well as clearing of forests and other land-use changes. Figure source: NOAA NCDC.

Detection and attribution of climate signals, as well as its common-sense meaning, has a more precise definition within the climate change literature, as expressed by the IPCC. Detection of a climate signal does not always imply significant attribution. The IPCC's Fourth Assessment Report says "it is *extremely likely* that human activities have exerted a substantial net warming influence on climate since 1750," where "extremely likely" indicates a probability greater than 95%. *Detection* of a signal requires demonstrating that an observed change is statistically significantly different from that which can be explained by natural internal variability.

Attribution requires demonstrating that a signal is:

- unlikely to be due entirely to internal variability;

- consistent with the estimated responses to the given combination of anthropogenic and natural forcing

- not consistent with alternative, physically plausible explanations of recent climate change that exclude important elements of the given combination of forcings.

Key Attributions

Greenhouse Gases

Carbon dioxide is the primary greenhouse gas that is contributing to recent climate change. CO_2 is absorbed and emitted naturally as part of the carbon cycle, through animal and plant respiration, volcanic eruptions, and ocean-atmosphere exchange. Human activities, such as the burning of fossil fuels and changes in land use, release large amounts of carbon to the atmosphere, causing CO_2 concentrations in the atmosphere to rise.

The high-accuracy measurements of atmospheric CO_2 concentration, initiated by Charles David Keeling in 1958, constitute the master time series documenting the changing composition of the atmosphere. These data have iconic status in climate change science as evidence of the effect of human activities on the chemical composition of the global atmosphere.

Along with CO_2, methane and nitrous oxide are also major forcing contributors to the greenhouse effect. The Kyoto Protocol lists these together with hydrofluorocarbons (HFCs), perfluorocarbons (PFCs), and sulphur hexafluoride (SF_6), which are entirely artificial (i.e. anthropogenic) gases, which also contribute to radiative forcing in the atmosphere. The chart at right attributes anthropogenic greenhouse gas emissions to eight main economic sectors, of which the largest contributors are power stations (many of which burn coal or other fossil fuels), industrial processes, transportation fuels (generally fossil fuels), and agricultural by-products (mainly methane from enteric fermentation and nitrous oxide from fertilizer use).

Water Vapor

Water vapor is the most abundant greenhouse gas and also the most important in terms of its contribution to the natural greenhouse effect, despite having a short atmospheric lifetime (about 10 days). Some human activities can influence local water vapor levels. However, on a global scale, the concentration of water vapor is controlled by temperature, which influences overall rates of evaporation and precipitation. Therefore, the global concentration of water vapor is not substantially affected by direct human emissions.

Annual Greenhouse Gas Emissions by Sector

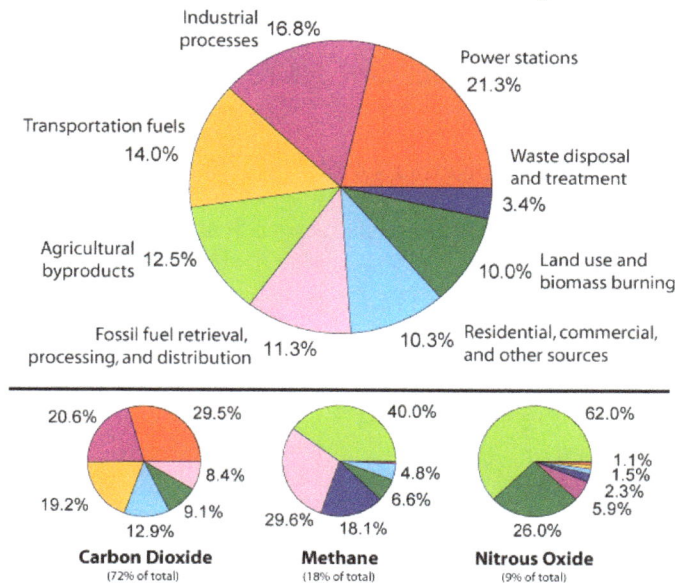

Industrial processes 16.8%
Power stations 21.3%
Transportation fuels 14.0%
Waste disposal and treatment 3.4%
Agricultural byproducts 12.5%
Land use and biomass burning 10.0%
Fossil fuel retrieval, processing, and distribution 11.3%
Residential, commercial, and other sources 10.3%

Carbon Dioxide (72% of total): 29.5%, 8.4%, 9.1%, 12.9%, 19.2%, 20.6%
Methane (18% of total): 40.0%, 4.8%, 6.6%, 18.1%, 29.6%
Nitrous Oxide (9% of total): 62.0%, 1.1%, 1.5%, 2.3%, 5.9%, 26.0%

Emission Database for Global Atmospheric Research version 3.2, fast track 2000 project

Land Use

Climate change is attributed to land use for two main reasons. Between 1750 and 2007, about two-thirds of anthropogenic CO_2 emissions were produced from burning fossil fuels, and about one-third of emissions from changes in land use, primarily deforestation. Deforestation both reduces the amount of carbon dioxide absorbed by deforested regions and releases greenhouse gases directly, together with aerosols, through biomass burning that frequently accompanies it.

A second reason that climate change has been attributed to land use is that the terrestrial albedo is often altered by use, which leads to radiative forcing. This effect is more significant locally than globally.

Livestock and Land Use

Worldwide, livestock production occupies 70% of all land used for agriculture, or 30% of the ice-free land surface of the Earth. More than 18% of anthropogenic greenhouse gas emissions are attributed to livestock and livestock-related activities such as deforestation and increasingly fuel-intensive farming practices. Specific attributions to the livestock sector include:

- 9% of global anthropogenic carbon dioxide emissions

- 35–40% of global anthropogenic methane emissions (chiefly due to enteric fermentation and manure)

- 64% of global anthropogenic nitrous oxide emissions, chiefly due to fertilizer use.

Aerosols

With virtual certainty, scientific consensus has attributed various forms of climate change, chiefly cooling effects, to aerosols, which are small particles or droplets suspended in the atmosphere. Key sources to which anthropogenic aerosols are attributed include:

- biomass burning such as slash and burn deforestation. Aerosols produced are primarily black carbon.

- industrial air pollution, which produces soot and airborne sulfates, nitrates, and ammonium

- dust produced by land use effects such as desertification

Attribution of 20th Century Climate Change

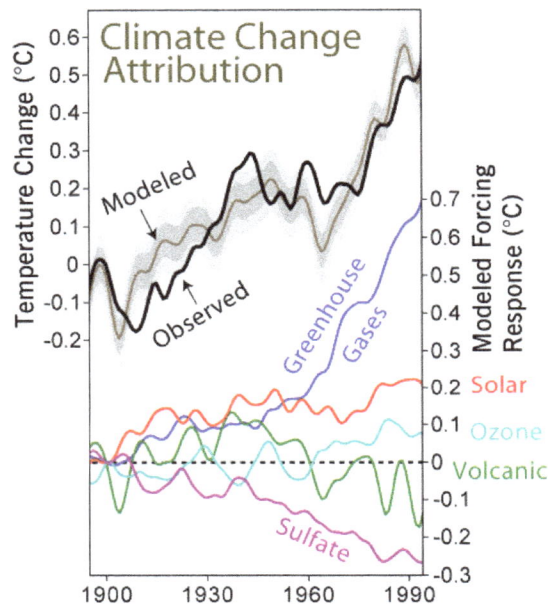

One global climate model's reconstruction of temperature change during the 20th century as the result of five studied forcing factors and the amount of temperature change attributed to each.

Over the past 150 years human activities have released increasing quantities of greenhouse gases into the atmosphere. This has led to increases in mean global temperature, or global warming. Other human effects are relevant—for example, sulphate aerosols are believed to have a cooling effect. Natural factors also contribute. According to the historical temperature record of the last century, the Earth's near-surface air temperature has risen around 0.74 ± 0.18 °Celsius (1.3 ± 0.32 °Fahrenheit).

A historically important question in climate change research has regarded the relative importance of human activity and non-anthropogenic causes during the period of instrumental record. In the 1995 Second Assessment Report (SAR), the IPCC made the widely quoted statement that "The balance of evidence suggests a discernible human influence on global climate". The phrase "balance of evidence" suggested the (English) common-law standard of proof required in civil as opposed to criminal courts: not as high as "beyond reasonable doubt". In 2001 the Third Assessment Report

(TAR) refined this, saying "There is new and stronger evidence that most of the warming observed over the last 50 years is attributable to human activities". The 2007 Fourth Assessment Report (AR4) strengthened this finding:

- "Anthropogenic warming of the climate system is widespread and can be detected in temperature observations taken at the surface, in the free atmosphere and in the oceans. Evidence of the effect of external influences, both anthropogenic and natural, on the climate system has continued to accumulate since the TAR."

Other findings of the IPCC Fourth Assessment Report include:

- "It is *extremely unlikely* (<5%) that the global pattern of warming during the past half century can be explained without external forcing (i.e., it is inconsistent with being the result of internal variability), and *very unlikely* that it is due to known natural external causes alone. The warming occurred in both the ocean and the atmosphere and took place at a time when natural external forcing factors would likely have produced cooling."

- "From new estimates of the combined anthropogenic forcing due to greenhouse gases, aerosols, and land surface changes, it is *extremely likely* (>95%) that human activities have exerted a substantial net warming influence on climate since 1750."

- "It is *virtually certain* that anthropogenic aerosols produce a net negative radiative forcing (cooling influence) with a greater magnitude in the Northern Hemisphere than in the Southern Hemisphere."

Over the past five decades there has been a global warming of approximately 0.65 °C (1.17 °F) at the Earth's surface. Among the possible factors that could pro-duce changes in global mean temperature are internal variability of the climate system, external forcing, an increase in concentration of greenhouse gases, or any combination of these. Current studies indicate that the increase in greenhouse gases, most notably CO_2, is mostly responsible for the observed warming. Evidence for this conclusion includes:

- Estimates of internal variability from climate models, and reconstructions of past temperatures, indicate that the warming is unlikely to be entirely natural.

- Climate models forced by natural factors *and* increased greenhouse gases and aerosols reproduce the observed global temperature changes; those forced by natural factors alone do not.

- "Fingerprint" methods indicate that the pattern of change is closer to that expected from greenhouse gas-forced change than from natural change.

- The plateau in warming from the 1940s to 1960s can be attributed largely to sulphate aerosol cooling.

Details on Attribution

Recent scientific assessments find that most of the warming of the Earth's surface over the past 50 years has been caused by human activities. This conclusion rests on multiple lines of evidence.

Like the warming "signal" that has gradually emerged from the "noise" of natural climate variability, the scientific evidence for a human influence on global climate has accumulated over the past several decades, from many hundreds of studies. No single study is a "smoking gun." Nor has any single study or combination of studies undermined the large body of evidence supporting the conclusion that human activity is the primary driver of recent warming.

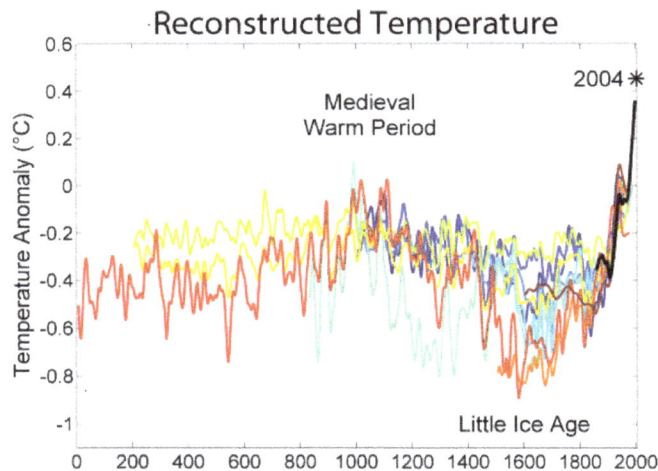

For Northern Hemisphere temperature, recent decades appear to be the warmest since at least about 1000AD, and the warming since the late 19th century is unprecedented over the last 1000 years. Older data are insufficient to provide reliable hemispheric temperature estimates.

The first line of evidence is based on a physical understanding of how greenhouse gases trap heat, how the climate system responds to increases in greenhouse gases, and how other human and natural factors influence climate. The second line of evidence is from indirect estimates of climate changes over the last 1,000 to 2,000 years. These records are obtained from living things and their remains (like tree rings and corals) and from physical quantities (like the ratio between lighter and heavier isotopes of oxygen in ice cores), which change in measurable ways as climate changes. The lesson from these data is that global surface temperatures over the last several decades are clearly unusual, in that they were higher than at any time during at least the past 400 years. For the Northern Hemisphere, the recent temperature rise is clearly unusual in at least the last 1,000 years.

The third line of evidence is based on the broad, qualitative consistency between observed changes in climate and the computer model simulations of how climate would be expected to change in response to human activities. For example, when climate models are run with historical increases in greenhouse gases, they show gradual warming of the Earth and ocean surface, increases in ocean heat content and the temperature of the lower atmosphere, a rise in global sea level, retreat of sea ice and snow cover, cooling of the stratosphere, an increase in the amount of atmospheric water vapor, and changes in large-scale precipitation and pressure patterns. These and other aspects of modelled climate change are in agreement with observations.

"Fingerprint" Studies

Reconstructions of global temperature that include greenhouse gas increases and other human influences (red line, based on many models) closely match measured temperatures (dashed line). Those that only include natural influences (blue line, based on many models) show a slight cooling, which has not occurred. The ability of models to generate reasonable histories of global temperature is verified by their response to four 20th-century volcanic eruptions: each eruption caused brief cooling that appeared in observed as well as modeled records.

Finally, there is extensive statistical evidence from so-called "fingerprint" studies. Each factor that affects climate produces a unique pattern of climate response, much as each person has a unique fingerprint. Fingerprint studies exploit these unique signatures, and allow detailed comparisons of modelled and observed climate change patterns. Scientists rely on such studies to attribute observed changes in climate to a particular cause or set of causes. In the real world, the climate changes that have occurred since the start of the Industrial Revolution are due to a complex mixture of human and natural causes. The importance of each individual influence in this mixture changes over time. Of course, there are not multiple Earths, which would allow an experimenter to change one factor at a time on each Earth, thus helping to isolate different fingerprints. Therefore, climate models are used to study how individual factors affect climate. For example, a single factor (like greenhouse gases) or a set of factors can be varied, and the response of the modelled climate system to these individual or combined changes can thus be studied.

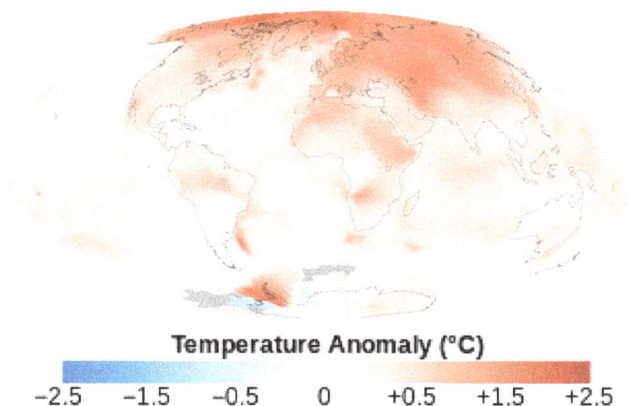

Temperature Anomaly (°C)

−2.5   −1.5   −0.5   0   +0.5   +1.5   +2.5

Two fingerprints of human activities on the climate are that land areas will warm more than the oceans, and that high latitudes will warm more than low latitudes. These projections have been confirmed by observations (shown above).

For example, when climate model simulations of the last century include all of the major influences on climate, both human-induced and natural, they can reproduce many important features of observed climate change patterns. When human influences are removed from the model experiments, results suggest that the surface of the Earth would actually have cooled slightly over the last 50 years. The clear message from fingerprint studies is that the observed warming over the last half-century cannot be explained by natural factors, and is instead caused primarily by human factors.

Another fingerprint of human effects on climate has been identified by looking at a slice through the layers of the atmosphere, and studying the pattern of temperature changes from the surface up through the stratosphere. The earliest fingerprint work focused on changes in surface and atmospheric temperature. Scientists then applied fingerprint methods to a whole range of climate variables, identifying human-caused climate signals in the heat content of the oceans, the height of the tropopause (the boundary between the troposphere and strato-sphere, which has shifted upward by hundreds of feet in recent decades), the geographical patterns of precipitation, drought, surface pressure, and the runoff from major river basins.

Studies published after the appearance of the IPCC Fourth Assessment Report in 2007 have also found human fingerprints in the increased levels of atmospheric moisture (both close to the surface and over the full extent of the atmosphere), in the decline of Arctic sea ice extent, and in the patterns of changes in Arctic and Antarctic surface temperatures.

The message from this entire body of work is that the climate system is telling a consistent story of increasingly dominant human influence – the changes in temperature, ice extent, moisture, and circulation patterns fit together in a physically consistent way, like pieces in a complex puzzle.

Increasingly, this type of fingerprint work is shifting its emphasis. As noted, clear and compelling scientific evidence supports the case for a pronounced human influence on global climate. Much of the recent attention is now on climate changes at continental and regional scales, and on variables that can have large impacts on societies. For example, scientists have established causal links between human activities and the changes in snowpack, maximum and minimum (diurnal) temperature, and the seasonal timing of runoff over mountainous regions of the western United States. Human activity is likely to have made a substantial contribution to ocean surface temperature changes in hurricane formation regions. Researchers are also looking beyond the physical climate system, and are beginning to tie changes in the distribution and seasonal behaviour of plant and animal species to human-caused changes in temperature and precipitation.

For over a decade, one aspect of the climate change story seemed to show a significant difference between models and observations. In the tropics, all models predicted that with a rise in greenhouse gases, the troposphere would be expected to warm more rapidly than the surface. Observations from weather balloons, satellites, and surface thermometers seemed to show the opposite behaviour (more rapid warming of the surface than the troposphere). This issue was a stumbling block in understanding the causes of climate change. It is now largely resolved. Research showed that there were large uncertainties in the satellite and weather balloon data. When uncertainties in models and observations are properly accounted for, newer observational data sets (with better treatment of known problems) are in agreement with climate model results.

This set of graphs shows the estimated contribution of various natural and human factors to changes in global mean temperature between 1889–2006. Estimated contributions are based on multivariate analysis rather than model simulations. The graphs show that human influence on climate has eclipsed the magnitude of natural temperature changes over the past 120 years. Natural influences on temperature—El Niño, solar variability, and volcanic aerosols—have varied approximately plus and minus 0.2 °C (0.4 °F), (averaging to about zero), while human influences have contributed roughly 0.8 °C (1 °F) of warming since 1889.

This does not mean, however, that all remaining differences between models and observations have been resolved. The observed changes in some climate variables, such as Arctic sea ice, some aspects of precipitation, and patterns of surface pressure, appear to be proceeding much more rapidly than models have projected. The reasons for these differences are not well understood. Nevertheless, the bottom-line conclusion from climate fingerprinting is that most of the observed changes studied to date are consistent with each other, and are also consistent with our scientific understanding of how the climate system would be expected to respond to the increase in heat-trapping gases resulting from human activities.

Extreme Weather Events

SHIFTING DISTRIBUTION OF SUMMER TEMPERATURE ANOMALIES

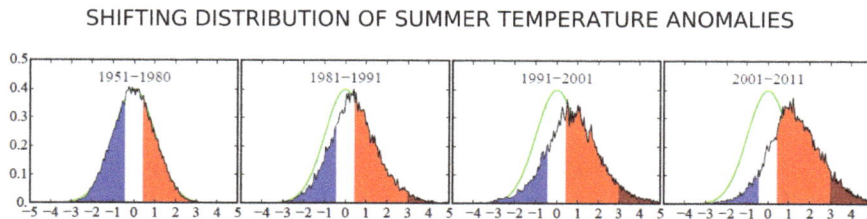

Credit: James Hansen, NASA Goddard Institute for Space Studies

Frequency of occurrence (vertical axis) of local June–July–August temperature anomalies (relative to 1951–1980 mean) for Northern Hemisphere land in units of local standard deviation (horizontal axis). According to Hansen *et al.* (2012), the distribution of anomalies has shifted to the right as a consequence of global warming, meaning that unusually hot summers have become more common. This is analogous to the rolling of a dice: cool summers now cover only half of one side of a six-sided die, white covers one side, red covers four sides, and an extremely hot (red-brown) anomaly covers half of one side.

One of the subjects discussed in the literature is whether or not extreme weather events can be attributed to human activities. Seneviratne *et al.* (2012) stated that attributing individual extreme weather events to human activities was challenging. They were, however, more confident over attributing changes in long-term trends of extreme weather. For example, Seneviratne *et al.* (2012) concluded that human activities had likely led to a warming of extreme daily minimum and maximum temperatures at the global scale.

Another way of viewing the problem is to consider the effects of human-induced climate change on the probability of future extreme weather events. Stott *et al.* (2003), for example, considered whether or not human activities had increased the risk of severe heat waves in Europe, like the one experienced in 2003. Their conclusion was that human activities had very likely more than doubled the risk of heat waves of this magnitude.

An analogy can be made between an athlete on steroids and human-induced climate change. In the same way that an athlete's performance may increase from using steroids, human-induced climate change increases the risk of some extreme weather events.

Hansen *et al.* (2012) suggested that human activities have greatly increased the risk of summertime heat waves. According to their analysis, the land area of the Earth affected by very hot summer temperature anomalies has greatly increased over time (refer to graphs on the left). In the base period 1951-1980, these anomalies covered a few tenths of 1% of the global land area. In recent years, this has increased to around 10% of the global land area. With high confidence, Hansen *et al.* (2012) attributed the 2010 Moscow and 2011 Texas heat waves to human-induced global warming.

An earlier study by Dole *et al.* (2011) concluded that the 2010 Moscow heatwave was mostly due to natural weather variability. While not directly citing Dole *et al.* (2011), Hansen *et al.* (2012) rejected this type of explanation. Hansen *et al.* (2012) stated that a combination of natural weather variability and human-induced global warming was responsible for the Moscow and Texas heat waves.

Scientific Literature and Opinion

There are a number of examples of published and informal support for the consensus view. As mentioned earlier, the IPCC has concluded that most of the observed increase in globally averaged temperatures since the mid-20th century is "very likely" due to human activities. The IPCC's conclusions are consistent with those of several reports produced by the US National Research Council. A report published in 2009 by the U.S. Global Change Research Program concluded that "[global] warming is unequivocal and primarily human-induced." A number of scientific organizations have issued statements that support the consensus view. Two examples include:

- a joint statement made in 2005 by the national science academies of the G8, and Brazil, China and India;

- a joint statement made in 2008 by the Network of African Science Academies.

Detection and Attribution Studies

The IPCC Fourth Assessment Report (2007), concluded that attribution was possible for a number of observed changes in the climate. However, attribution was found to be more difficult when assessing changes over smaller regions (less than continental scale) and over short time periods (less than 50 years). Over larger regions, averaging reduces natural variability of the climate, making detection and attribution easier.

- In 1996, in a paper in *Nature* titled "A search for human influences on the thermal structure of the atmosphere", Benjamin D. Santer et al. wrote: "The observed spatial patterns of temperature change in the free atmosphere from 1963 to 1987 are similar to those predicted by state-of-the-art climate models incorporating various combinations of changes in carbon dioxide, anthropogenic sulphate aerosol and stratospheric ozone concentrations. The degree of pattern similarity between models and observations increases through this period. It is likely that this trend is partially due to human activities, although many uncertainties remain, particularly relating to estimates of natural variability."

- A 2002 paper in the *Journal of Geophysical Research* says "Our analysis suggests that the early twentieth century warming can best be explained by a combination of warming due to increases in greenhouse gases and natural forcing, some cooling due to other anthropogenic forcings, and a substantial, but not implausible, contribution from internal variability. In the second half of the century we find that the warming is largely caused by changes in greenhouse gases, with changes in sulphates and, perhaps, volcanic aerosol offsetting approximately one third of the warming."

- A 2005 review of detection and attribution studies by the International Ad Hoc Detection and Attribution Group found that "natural drivers such as solar variability and volcanic activity are at most partially responsible for the large-scale temperature changes observed over the past century, and that a large fraction of the warming over the last 50 yr can be attributed to greenhouse gas increases. Thus, the recent research supports and strengthens the IPCC Third Assessment Report conclusion that 'most of the global warming over the past 50 years is likely due to the increase in greenhouse gases.'"

- Barnett and colleagues (2005) say that the observed warming of the oceans "cannot be explained by natural internal climate variability or solar and volcanic forcing, but is well simulated by two anthropogenically forced climate models," concluding that "it is of human origin, a conclusion robust to observational sampling and model differences".

- Two papers in the journal *Science* in August 2005 resolve the problem, evident at the time of the TAR, of tropospheric temperature trends. The UAH version of the record contained errors, and there is evidence of spurious cooling trends in the radiosonde record, particularly in the tropics. See satellite temperature measurements for details; and the 2006 US CCSP report.

- Multiple independent reconstructions of the temperature record of the past 1000 years confirm that the late 20th century is probably the warmest period in that time

Reviews of Scientific Opinion

- An essay in *Science* surveyed 928 abstracts related to climate change, and concluded that most journal reports accepted the consensus. This is discussed further in scientific opinion on climate change.

- A 2010 paper in the Proceedings of the National Academy of Sciences found that among a pool of roughly 1,000 researchers who work directly on climate issues and publish the most frequently on the subject, 97% agree that anthropogenic climate change is happening.

- A 2011 paper from George Mason University published in the *International Journal of Public Opinion Research*, "The Structure of Scientific Opinion on Climate Change," collected the opinions of scientists in the earth, space, atmospheric, oceanic or hydrological sciences. The 489 survey respondents—representing nearly half of all those eligible according to the survey's specific standards – work in academia, government, and industry, and are members of prominent professional organizations. The study found that 97% of the 489 scientists surveyed agreed that global temperatures have risen over the past century. Moreover, 84% agreed that "human-induced greenhouse warming" is now occurring." Only 5% disagreed with the idea that human activity is a significant cause of global warming.

As described above, a small minority of scientists do disagree with the consensus: see list of scientists opposing global warming consensus. For example, Willie Soon and Richard Lindzen say that there is insufficient proof for anthropogenic attribution. Generally this position requires new physical mechanisms to explain the observed warming.

Solar Activity

Solar radiation at the top of our atmosphere, and global temperature

Modelled simulation of the effect of various factors (including GHGs, Solar irradiance) singly and in combination, showing in particular that solar activity produces a small and nearly uniform warming, unlike what is observed.

Solar sunspot maximum occurs when the magnetic field of the sun collapses and reverse as part of its average 11 year solar cycle (22 years for complete North to North restoration).

The role of the sun in recent climate change has been looked at by climate scientists. Since 1978, output from the Sun has been measured by satellites significantly more accurately than was previously possible from the surface. These measurements indicate that the Sun's total solar irradiance has not increased since 1978, so the warming during the past 30 years cannot be directly attributed to an increase in total solar energy reaching the Earth. In the three decades since 1978, the combination of solar and volcanic activity probably had a slight cooling influence on the climate.

Climate models have been used to examine the role of the sun in recent climate change. Models are unable to reproduce the rapid warming observed in recent decades when they only take into account variations in total solar irradiance and volcanic activity. Models are, however, able to simulate the observed 20th century changes in temperature when they include all of the most important external forcings, including human influences and natural forcings. As has already been stated, Hegerl *et al.* (2007) concluded that greenhouse gas forcing had "very likely" caused most of the observed global warming since the mid-20th century. In making this conclusion, Hegerl *et al.* (2007) allowed for the possibility that climate models had been underestimated the effect of solar forcing.

The role of solar activity in climate change has also been calculated over longer time periods using "proxy" datasets, such as tree rings. Models indicate that solar and volcanic forcings can explain periods of relative warmth and cold between A.D. 1000 and 1900, but human-induced forcings are needed to reproduce the late-20th century warming.

Another line of evidence against the sun having caused recent climate change comes from looking at how temperatures at different levels in the Earth's atmosphere have changed. Models and observations show that greenhouse gas results in warming of the lower atmosphere at the surface (called the troposphere) but cooling of the upper atmosphere (called the stratosphere). Depletion of the ozone layer by chemical refrigerants has also resulted in a cooling effect in the stratosphere. If the sun was responsible for observed warming, warming of the troposphere at the surface and warming at the top of the stratosphere would be expected as increase solar activity would replenish ozone and oxides of nitrogen. The stratosphere has a reverse temperature gradient than the troposphere so as the temperature of the troposphere cools with altitude, the stratosphere rises with altitude. Hadley cells are the mechanism by which equatorial generated ozone in the tropics (highest area of UV irradiance in the stratosphere) is moved poleward. Global climate models suggest that climate change may widen the Hadley cells and push the jetstream northward thereby expanding the tropics region and resulting in warmer, dryer conditions in those areas overall.

Non-consensus Views

Habibullo Abdussamatov (2004), head of space research at St. Petersburg's Pulkovo Astronomical Observatory in Russia, has argued that the sun is responsible for recently observed climate change. Journalists for news sources canada.com (Solomon, 2007b), National Geographic News (Ravillious, 2007), and LiveScience (Than, 2007) reported on the story of warming on Mars. In these articles, Abdussamatov was quoted. He stated that warming on Mars was evidence that global warming on Earth was being caused by changes in the sun.

Radiative forcing components

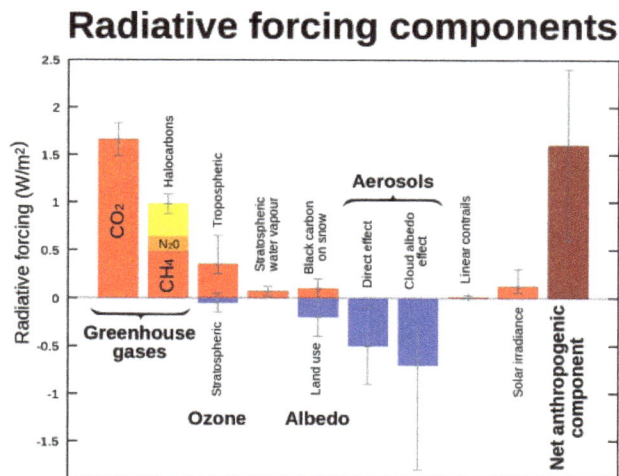

Contribution of natural factors and human activities to radiative forcing of climate change. Radiative forcing values are for the year 2005, relative to the pre-industrial era (1750). The contribution of solar irradiance to radiative forcing is 5% the value of the combined radiative forcing due to increases in the atmospheric concentrations of carbon dioxide, methane and nitrous oxide.

Ravillious (2007) quoted two scientists who disagreed with Abdussamatov: Amato Evan, a climate scientist at the University of Wisconsin-Madison, in the US, and Colin Wilson, a planetary physicist at Oxford University in the UK. According to Wilson, "Wobbles in the orbit of Mars are the main cause of its climate change in the current era". Than (2007) quoted Charles Long, a climate physicist at Pacific Northwest National Laboratories in the US, who disagreed with Abdussamatov.

Than (2007) pointed to the view of Benny Peiser, a social anthropologist at Liverpool John Moores University in the UK. In his newsletter, Peiser had cited a blog that had commented on warming observed on several planetary bodies in the Solar system. These included Neptune's moon Triton, Jupiter, Pluto and Mars. In an e-mail interview with Than (2007), Peiser stated that:

"I think it is an intriguing coincidence that warming trends have been observed on a number of very diverse planetary bodies in our solar system, (...) Perhaps this is just a fluke."

Than (2007) provided alternative explanations of why warming had occurred on Triton, Pluto, Jupiter and Mars.

The US Environmental Protection Agency (US EPA, 2009) responded to public comments on climate change attribution. A number of commenters had argued that recent climate change could be attributed to changes in solar irradiance. According to the US EPA (2009), this attribution was not supported by the bulk of the scientific literature. Citing the work of the IPCC (2007), the US EPA pointed to the low contribution of solar irradiance to radiative forcing since the start of the Industrial Revolution in 1750. Over this time period (1750 to 2005), the estimated contribution of solar irradiance to radiative forcing was 5% the value of the combined radiative forcing due to increases in the atmospheric concentrations of carbon dioxide, methane and nitrous oxide.

Effect of Cosmic Rays

Henrik Svensmark has suggested that the magnetic activity of the sun deflects cosmic rays, and that this may influence the generation of cloud condensation nuclei, and thereby have an effect on the climate. The website ScienceDaily reported on a 2009 study that looked at how past changes in climate have been affected by the Earth's magnetic field. Geophysicist Mads Faurschou Knudsen, who co-authored the study, stated that the study's results supported Svensmark's theory. The authors of the study also acknowledged that CO_2 plays an important role in climate change.

Consensus View on cosmic Rays

The view that cosmic rays could provide the mechanism by which changes in solar activity affect climate is not supported by the literature. Solomon *et al.* (2007) state:

"[..] the cosmic ray time series does not appear to correspond to global total cloud cover after 1991 or to global low-level cloud cover after 1994. Together with the lack of a proven physical mechanism and the plausibility of other causal factors affecting changes in cloud cover, this makes the association between galactic cosmic ray-induced changes in aerosol and cloud formation controversial."

Studies by Lockwood and Fröhlich (2007) and Sloan and Wolfendale (2008) found no relation between warming in recent decades and cosmic rays. Pierce and Adams (2009) used a model to simulate the effect of cosmic rays on cloud properties. They concluded that the hypothesized effect of cosmic rays was too small to explain recent climate change. Pierce and Adams (2009) noted that their findings did not rule out a possible connection between cosmic rays and climate change, and recommended further research.

Erlykin *et al.* (2009) found that the evidence showed that connections between solar variation and climate were more likely to be mediated by direct variation of insolation rather than cosmic rays, and concluded: "Hence within our assumptions, the effect of varying solar activity, either by direct solar irradiance or by varying cosmic ray rates, must be less than 0.07 °C since 1956, i.e. less than 14% of the observed global warming." Carslaw (2009) and Pittock (2009) review the recent and historical literature in this field and continue to find that the link between cosmic rays and climate is tenuous, though they encourage continued research. US EPA (2009) commented on research by Duplissy *et al.* (2009):

The CLOUD experiments at CERN are interesting research but do not provide conclusive evidence that cosmic rays can serve as a major source of cloud seeding. Preliminary results from the experiment (Duplissy et al., 2009) suggest that though there was some evidence of ion mediated nucleation, for most of the nucleation events observed the contribution of ion processes appeared to be minor. These experiments also showed the difficulty in maintaining sufficiently clean conditions and stable temperatures to prevent spurious aerosol bursts. There is no indication that the earlier Svensmark experiments could even have matched the controlled conditions of the CERN experiment. We find that the Svensmark results on cloud seeding have not yet been shown to be robust or sufficient to materially alter the conclusions of the assessment literature, especially given the abundance of recent literature that is skeptical of the cosmic ray-climate linkage.

Greenhouse Effect

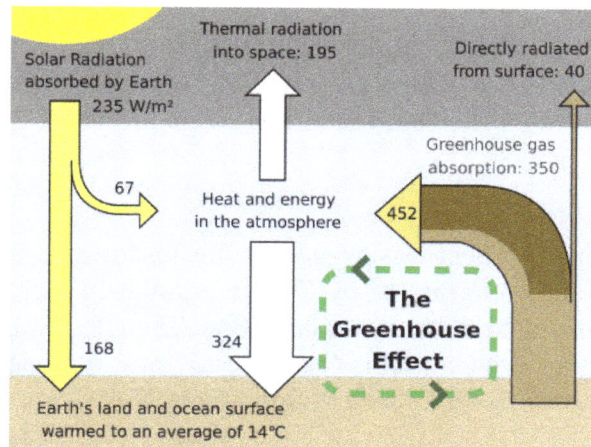

Thermal radiation into space: 195

Solar Radiation absorbed by Earth 235 W/m²

Directly radiated from surface: 40

Greenhouse gas absorption: 350

67

Heat and energy in the atmosphere

452

The Greenhouse Effect

168

324

Earth's land and ocean surface warmed to an average of 14°C

A representation of the exchanges of energy between the source (the Sun), Earth's surface, the Earth's atmosphere, and the ultimate sink outer space. The ability of the atmosphere to capture and recycle energy emitted by Earth's surface is the defining characteristic of the greenhouse effect.

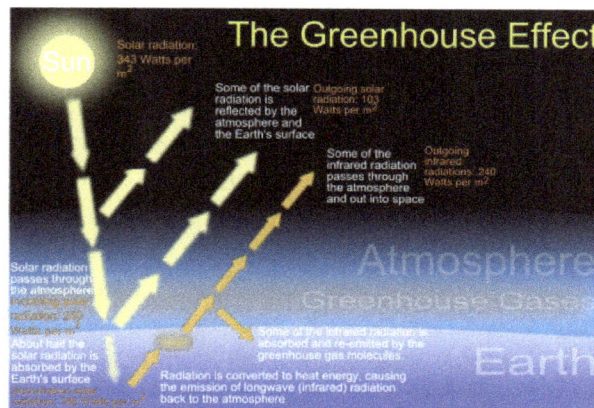

Another diagram of the greenhouse effect

The greenhouse effect is the process by which radiation from a planet's atmosphere warms the planet's surface to a temperature above what it would be without its atmosphere.

If a planet's atmosphere contains radiatively active gases (i.e., greenhouse gases) the atmosphere will radiate energy in all directions. Part of this radiation is directed towards the surface, warming it. The downward component of this radiation – that is, the strength of the greenhouse effect – will depend on the atmosphere's temperature and on the amount of greenhouse gases that the atmosphere contains.

On Earth, the atmosphere is warmed by absorption of infrared thermal radiation from the underlying surface, absorption of shorter wavelength radiant energy from the sun, and convective heat fluxes from the surface. Greenhouse gases in the atmosphere radiate energy, some of which is directed to the surface and lower atmosphere. The mechanism that produces this difference between the actual surface temperature and the effective temperature is due to the atmosphere and is known as the greenhouse effect.

Earth's natural greenhouse effect is critical to supporting life. Human activities, primarily the burning of fossil fuels and clearing of forests, have intensified the natural greenhouse effect, causing global warming.

The mechanism is named after a faulty analogy with the effect of solar radiation passing through glass and warming a greenhouse. The way a greenhouse retains heat is fundamentally different, as a greenhouse works by reducing airflow and retaining warm air inside the structure.

History

The existence of the greenhouse effect was argued for by Joseph Fourier in 1824. The argument and the evidence was further strengthened by Claude Pouillet in 1827 and 1838, and reasoned from experimental observations by John Tyndall in 1859. The effect was more fully quantified by Svante Arrhenius in 1896. However, the term "greenhouse" was not used to refer to this effect by any of these scientists; the term was first used in this way by Nils Gustaf Ekholm in 1901.

In 1917 Alexander Graham Bell wrote "[The unchecked burning of fossil fuels] would have a sort of greenhouse effect", and "The net result is the greenhouse becomes a sort of hot-house." Bell went on to also advocate the use of alternate energy sources, such as solar energy.

Mechanism

Earth receives energy from the Sun in the form of ultraviolet, visible, and near-infrared radiation. Of the total amount of solar energy available at the top of the atmosphere, about 26% is reflected to space by the atmosphere and clouds and 19% is absorbed by the atmosphere and clouds. Most of the remaining energy is absorbed at the surface of Earth. Because the Earth's surface is colder than the photosphere of the Sun, it radiates at wavelengths that are much longer than the wavelengths that were absorbed. Most of this thermal radiation is absorbed by the atmosphere, thereby warming it. In addition to the absorption of solar and thermal radiation, the atmosphere further gains heat by sensible and latent heat fluxes from the surface. The atmosphere radiates energy both upwards and downwards; the part radiated downwards is absorbed by the surface of Earth. This leads to a higher equilibrium temperature than if the atmosphere were absent.

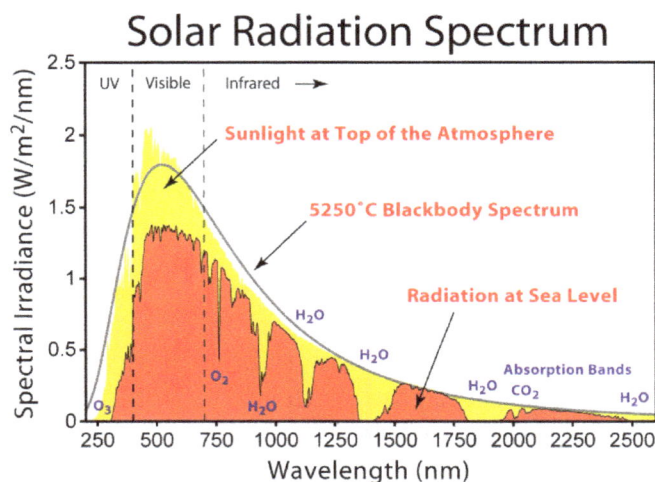

The solar radiation spectrum for direct light at both the top of Earth's atmosphere and at sea level

An ideal thermally conductive blackbody at the same distance from the Sun as Earth would have a temperature of about 5.3 °C. However, because Earth reflects about 30% of the incoming sunlight, this idealized planet's effective temperature (the temperature of a blackbody that would emit the same amount of radiation) would be about −18 °C. The surface temperature of this hypothetical planet is 33 °C below Earth's actual surface temperature of approximately 14 °C.

The basic mechanism can be qualified in a number of ways, none of which affect the fundamental process. The atmosphere near the surface is largely opaque to thermal radiation (with important exceptions for "window" bands), and most heat loss from the surface is by sensible heat and latent heat transport. Radiative energy losses become increasingly important higher in the atmosphere, largely because of the decreasing concentration of water vapor, an important greenhouse gas. It is more realistic to think of the greenhouse effect as applying to a "surface" in the mid-troposphere, which is effectively coupled to the surface by a lapse rate. The simple picture also assumes a steady state, but in the real world there are variations due to the diurnal cycle as well as the seasonal cycle and weather disturbances. Solar heating only applies during daytime. During the night, the atmosphere cools somewhat, but not greatly, because its emissivity is low. Diurnal temperature changes decrease with height in the atmosphere.

Within the region where radiative effects are important, the description given by the idealized greenhouse model becomes realistic. Earth's surface, warmed to a temperature around 255 K, radiates long-wavelength, infrared heat in the range of 4–100 μm. At these wavelengths, greenhouse gases that were largely transparent to incoming solar radiation are more absorbent. Each layer of atmosphere with greenhouses gases absorbs some of the heat being radiated upwards from lower layers. It reradiates in all directions, both upwards and downwards; in equilibrium (by definition) the same amount as it has absorbed. This results in more warmth below. Increasing the concentration of the gases increases the amount of absorption and reradiation, and thereby further warms the layers and ultimately the surface below.

Greenhouse gases—including most diatomic gases with two different atoms (such as carbon monoxide, CO) and all gases with three or more atoms—are able to absorb and emit infrared radiation. Though more than 99% of the dry atmosphere is IR transparent (because the main constituents—N_2, O_2, and Ar—are not able to directly absorb or emit infrared radiation), intermolecular collisions cause the energy absorbed and emitted by the greenhouse gases to be shared with the other, non-IR-active, gases.

Greenhouse Gases

Atmospheric gases only absorb some wavelengths of energy but are transparent to others. The absorption patterns of water vapor (blue peaks) and carbon dioxide (pink peaks) overlap in some wavelengths. Carbon dioxide is not as strong a greenhouse gas as water vapor, but it absorbs energy in wavelengths (12-15 micrometers) that water vapor does not, partially closing the "window" through which heat radiated by the surface would normally escape to space. (Illustration NASA, Robert Rohde)

By their percentage contribution to the greenhouse effect on Earth the four major gases are:

- water vapor, 36–70%

- carbon dioxide, 9–26%

- methane, 4–9%

- ozone, 3–7%

It is not physically realistic to assign a specific percentage to each gas because the absorption and emission bands of the gases overlap (hence the ranges given above). The major non-gas contributor to Earth's greenhouse effect, clouds, also absorb and emit infrared radiation and thus have an effect on the radiative properties of the atmosphere.

Role in Climate Change

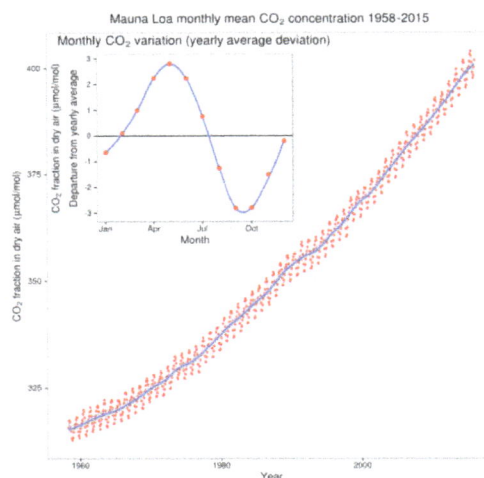

The Keeling Curve of atmospheric CO_2 concentrations measured at Mauna Loa Observatory.

Strengthening of the greenhouse effect through human activities is known as the enhanced (or anthropogenic) greenhouse effect. This increase in radiative forcing from human activity is attributable mainly to increased atmospheric carbon dioxide levels. According to the latest Assessment Report from the Intergovernmental Panel on Climate Change, "*atmospheric concentrations of carbon dioxide, methane and nitrous oxide are unprecedented in at least the last 800,000 years. Their effects, together with those of other anthropogenic drivers, have been detected throughout the climate system and are extremely likely to have been the dominant cause of the observed warming since the mid-20th century*".

CO_2 is produced by fossil fuel burning and other activities such as cement production and tropical deforestation. Measurements of CO_2 from the Mauna Loa observatory show that concentrations have increased from about 313 parts per million (ppm) in 1960 to about 389 ppm in 2010. It reached the 400 ppm milestone on May 9, 2013. The current observed amount of CO_2 exceeds the geological record maxima (~300 ppm) from ice core data. The effect of combustion-produced carbon dioxide on the global climate, a special case of the greenhouse effect first described in 1896 by Svante Arrhenius, has also been called the Callendar effect.

Over the past 800,000 years, ice core data shows that carbon dioxide has varied from values as low as 180 ppm to the pre-industrial level of 270 ppm. Paleoclimatologists consider variations in carbon dioxide concentration to be a fundamental factor influencing climate variations over this time scale.

Real Greenhouses

A modern Greenhouse in RHS Wisley

The "greenhouse effect" of the atmosphere is named by analogy to greenhouses which become warmer in sunlight. The explanation given in most sources for the warmer temperature in an actual greenhouse is that incident solar radiation in the visible, long-wavelength ultraviolet, and short-wavelength infrared range of the spectrum passes through the glass roof and walls and is absorbed by the floor, earth, and contents, which become warmer and re-emit the energy as longer-wavelength infrared radiation. Glass and other materials used for greenhouse walls do not transmit infrared radiation, so the infrared cannot escape via radiative transfer. As the structure is not open to the atmosphere, heat also cannot escape via convection, so the temperature inside the greenhouse rises. The greenhouse effect, due to infrared-opaque "greenhouse gases" including carbon dioxide and methane instead of glass, also affects Earth as a whole; there is no convective cooling because no significant amount of air escapes from Earth.

However the mechanism by which the atmosphere retains heat—the "greenhouse effect"—is different; a greenhouse is not primarily warmed by the "greenhouse effect". A greenhouse works primarily by allowing sunlight to warm surfaces inside the structure, but then preventing absorbed heat from leaving the structure through convection. The "greenhouse effect" heats Earth because greenhouse gases absorb outgoing radiative energy, heating the atmosphere which then emits radiative energy with some of it going back towards Earth.

A greenhouse is built of any material that passes sunlight, usually glass, or plastic. It mainly warms up because the sun warms the ground and contents inside, which then warms the air in the greenhouse. The air continues to heat up because it is confined within the greenhouse, unlike the environment outside the greenhouse where warm air near the surface rises and mixes with cooler air aloft. This can be demonstrated by opening a small window near the roof of a greenhouse: the temperature will drop considerably. It was demonstrated experimentally (R. W. Wood, 1909) that a "greenhouse" with a cover of rock salt (which is transparent to infrared) heats up an enclosure similarly to one with a glass cover. Thus greenhouses work primarily by preventing convective cooling.

More recent quantitative studies suggest that the effect of infrared radiative cooling is not negligibly small, and may have economic implications in a heated greenhouse. Analysis of issues of near-infrared radiation in a greenhouse with screens of a high coefficient of reflection concluded that installation of such screens reduced heat demand by about 8%, and application of dyes to transparent surfaces was suggested. Composite less-reflective glass, or less effective but cheaper anti-reflective coated simple glass, also produced savings.

Bodies other than Earth

In the Solar System, there also greenhouse effects on Mars, Venus, and Titan. The greenhouse effect on Venus is particularly large because its dense atmosphere consisting mainly of carbon dioxide. Titan has an anti-greenhouse effect, in that its atmosphere absorbs solar radiation but is relatively transparent to infrared radiation. Pluto is also colder than would be expected, because evaporation of nitrogen cools it.

A runaway greenhouse effect occurs if positive feedbacks lead to the evaporation of all greenhouse gases into the atmosphere. A runaway greenhouse effect involving carbon dioxide and water vapor is thought to have occurred on Venus.

Radiative Forcing

Radiative forcing or climate forcing is defined as the difference of insolation (sunlight) absorbed by the Earth and energy radiated back to space. Typically, radiative forcing is quantified at the tropopause in units of watts per square meter of the Earth's surface. A positive forcing (more incoming energy) warms the system, while negative forcing (more outgoing energy) cools it. Causes of radiative forcing include changes in insolation and the concentrations of radiatively active gases, commonly known as greenhouse gases and aerosols.

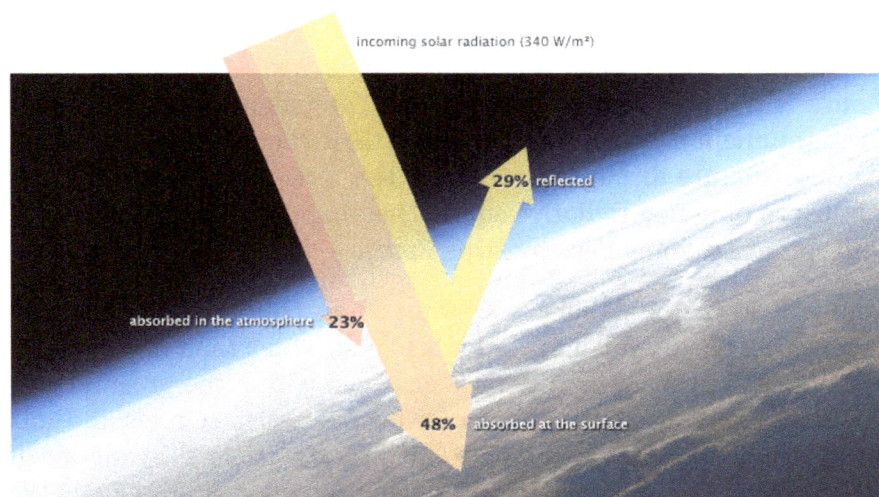

incoming solar radiation (340 W/m²)

29% reflected

absorbed in the atmosphere 23%

48% absorbed at the surface

Incoming solar radiation

Radiation Balance

Atmospheric gases only absorb some wavelengths of energy but are transparent to others. The absorption patterns of water vapor (blue peaks) and carbon dioxide (pink peaks) overlap in some wavelengths. Carbon dioxide is not as strong a greenhouse gas as water vapor, but it absorbs energy in wavelengths (12-15 micrometers) that water vapor does not, partially closing the "window" through which heat radiated by the surface would normally escape to space. (Illustration NASA, Robert Rohde)

Almost all of the energy that affects Earth's climate is received as radiant energy from the Sun. The planet and its atmosphere absorb and reflect some of the energy, while long-wave energy is radiated back into space. The balance between absorbed and radiated energy determines the average global temperature. Because the atmosphere absorbs some of the re-radiated long-wave energy, the planet is warmer than it would be in the absence of the atmosphere.

The radiation balance is altered by such factors as the intensity of solar energy, reflectivity of clouds or gases, absorption by various greenhouse gases or surfaces and heat emission by various materials. Any such alteration is a radiative forcing, and changes the balance. This happens continuously as sunlight hits the surface, clouds and aerosols form, the concentrations of atmospheric gases vary and seasons alter the ground cover.

IPCC Usage

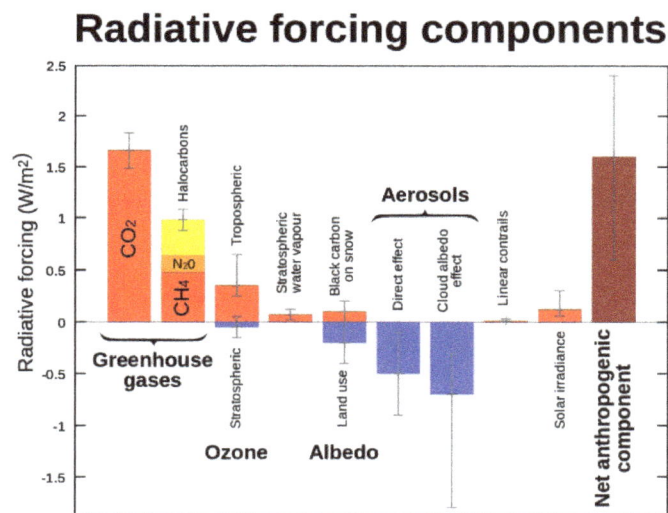

Radiative forcings, IPCC 2007.

The Intergovernmental Panel on Climate Change (IPCC) AR4 report defines radiative forcings as:

"Radiative forcing is a measure of the influence a factor has in altering the balance of incoming and outgoing energy in the Earth-atmosphere system and is an index of the importance of the factor as a potential climate change mechanism. In this report radiative forcing values are for changes relative to preindustrial conditions defined at 1750 and are expressed in Watts per square meter (W/m²)."

In simple terms, radiative forcing is "...the rate of energy change per unit area of the globe as measured at the top of the atmosphere." In the context of climate change, the term "forcing" is restricted to changes in the radiation balance of the surface-troposphere system imposed by external factors, with no changes in stratospheric dynamics, no surface and tropospheric feedbacks in operation (*i.e.*, no secondary effects induced because of changes in tropospheric motions or its thermodynamic state), and no dynamically induced changes in the amount and distribution of atmospheric water (vapour, liquid, and solid forms).

Climate Sensitivity

Radiative forcing can be used to estimate a subsequent change in equilibrium surface temperature (ΔT_s) arising from that forcing via the equation:

$$\Delta T_s = \lambda \Delta F$$

where λ is the climate sensitivity, usually with units in K/(W/m²), and ΔF is the radiative forcing. A typical value of λ is 0.8 K/(W/m²), which gives a warming of 3K for doubling of CO_2.

Example Calculations

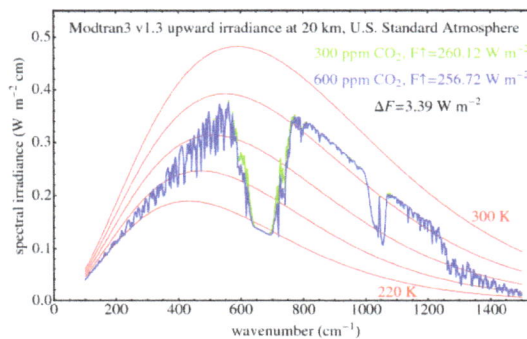

Radiative forcing for doubling CO_2, as calculated by radiative transfer code Modtran. Red lines are Planck curves.

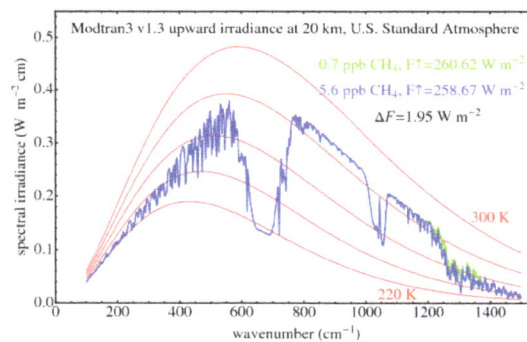

Radiative forcing for eight times increase of CH_4, as calculated by radiative transfer code Modtran.

Solar Forcing

Radiative forcing (measured in Watts per square meter) can be estimated in different ways for different components. For solar irradiance (*i.e.*, "solar forcing"), the radiative forcing is simply the change in the average amount of solar energy absorbed per square meter of the Earth's area. Since the Earth's cross-sectional area exposed to the Sun (πr^2) is equal to 1/4 of the surface area of the Earth ($4\pi r^2$), the solar input per unit area is one quarter the change in solar intensity. This must be multiplied by the fraction of incident sunlight that is absorbed, F=(1-R), where R is the reflectivity (albedo), of the Earth. The albedo is approximately 0.3, so F is approximately equal to 0.7. Thus, the solar forcing is the change in the solar intensity divided by 4 and multiplied by 0.7.

Likewise, a change in albedo will produce a solar forcing equal to the change in albedo divided by 4 multiplied by the solar constant.

Forcing due to Atmospheric Gas

For a greenhouse gas, such as carbon dioxide, radiative transfer codes that examine each spectral line for atmospheric conditions can be used to calculate the change ΔF as a function of changing concentration. These calculations can often be simplified into an algebraic formulation that is specific to that gas.

For instance, the simplified first-order approximation expression for carbon dioxide is:

$$\Delta F = 5.35 \times \ln \frac{C}{C_0} \, \mathrm{W \, m^{-2}}$$

where C is the CO_2 concentration in parts per million by volume and C_0 is the reference concentration. The relationship between carbon dioxide and radiative forcing is logarithmic and thus increased concentrations have a progressively smaller warming effect.

A different formula applies for other greenhouse gases such as methane and N_2O (square-root dependence) or CFCs (linear), with coefficients that can be found *e.g.* in the IPCC reports.

Related Measures

Radiative forcing is a useful way to compare different causes of perturbations in a climate system. Other possible tools can be constructed for the same purpose: for example Shine *et al.* say "... recent experiments indicate that for changes in absorbing aerosols and ozone, the predictive ability of radiative forcing is much worse... we propose an alternative, the 'adjusted troposphere and stratosphere forcing'. We present GCM calculations showing that it is a significantly more reliable predictor of this GCM's surface temperature change than radiative forcing. It is a candidate to supplement radiative forcing as a metric for comparing different mechanisms...". In this quote, GCM stands for "global circulation model", and the word "predictive" does not refer to the ability of GCMs to forecast climate change. Instead, it refers to the ability of the alternative tool proposed by the authors to help explain the system response.

History

The table below (derived from atmospheric radiative transfer models) shows changes in radiative forcing between 1979 and 2013. The table includes the contribution to radiative forcing from carbon dioxide (CO_2), methane (CH4), nitrous oxide (N2O); chlorofluorocarbons (CFCs) 12 and 11; and fifteen other minor, long-lived, halogenated gases. The table includes the contribution to radiative forcing of long-lived greenhouse gases. It does not include other forcings, such as aerosols and changes in solar activity.

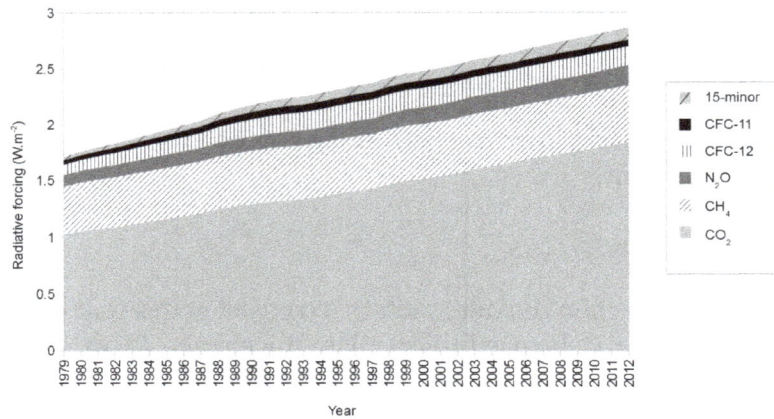

Changes in radiative forcing of long-lived greenhouse gases between 1979 and 2012.

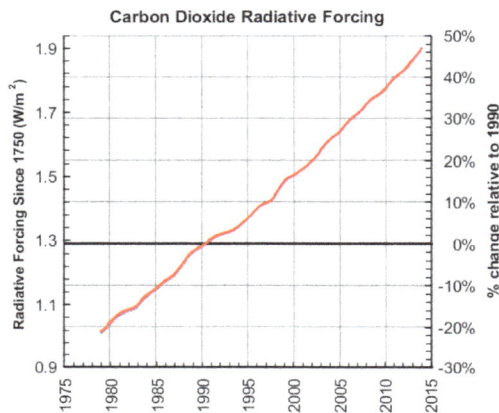

Radiative forcing, relative to 1750, due to carbon dioxide alone since 1979. The percent change from January 1, 1990 is shown on the right axis.

Global radiative forcing (relative to 1750, in $W\ m^{-2}$), CO_2-equivalent mixing ratio, and the Annual Greenhouse Gas Index (AGGI) between 1979-2014										
Year	CO_2	CH4	N2O	CFC-12	CFC-11	15-minor	Total	CO_2-eq ppm	AGGI 1990 = 1	AGGI % change
1979	1.027	0.419	0.104	0.092	0.039	0.031	1.712	383	0.786	
1980	1.058	0.426	0.104	0.097	0.042	0.034	1.761	386	0.808	2.8
1981	1.077	0.433	0.107	0.102	0.044	0.036	1.799	389	0.826	2.2
1982	1.089	0.440	0.111	0.108	0.046	0.038	1.831	391	0.841	1.8
1983	1.115	0.443	0.113	0.113	0.048	0.041	1.873	395	0.860	2.2
1984	1.140	0.446	0.116	0.118	0.050	0.044	1.913	397	0.878	2.2
1985	1.162	0.451	0.118	0.123	0.053	0.047	1.953	401	0.897	2.1
1986	1.184	0.456	0.122	0.129	0.056	0.049	1.996	404	0.916	2.2
1987	1.211	0.460	0.120	0.135	0.059	0.053	2.039	407	0.936	2.2
1988	1.250	0.464	0.123	0.143	0.062	0.057	2.099	412	0.964	3.0

1989	1.274	0.468	0.126	0.149	0.064	0.061	2.144	415	0.984	2.1
1990	1.293	0.472	0.129	0.154	0.065	0.065	2.178	418	1.000	1.6
1991	1.313	0.476	0.131	0.158	0.067	0.069	2.213	420	1.016	1.6
1992	1.324	0.480	0.133	0.162	0.067	0.072	2.238	422	1.027	1.1
1993	1.334	0.481	0.134	0.164	0.068	0.074	2.254	424	1.035	0.7
1994	1.356	0.483	0.134	0.166	0.068	0.075	2.282	426	1.048	1.3
1995	1.383	0.485	0.136	0.168	0.067	0.077	2.317	429	1.064	1.5
1996	1.410	0.486	0.139	0.169	0.067	0.078	2.350	431	1.079	1.4
1997	1.426	0.487	0.142	0.171	0.067	0.079	2.372	433	1.089	0.9
1998	1.465	0.491	0.145	0.172	0.067	0.080	2.419	437	1.111	2.0
1999	1.495	0.494	0.148	0.173	0.066	0.082	2.458	440	1.128	1.6
2000	1.513	0.494	0.151	0.173	0.066	0.083	2.481	442	1.139	0.9
2001	1.535	0.494	0.153	0.174	0.065	0.085	2.506	444	1.150	1.0
2002	1.564	0.494	0.156	0.174	0.065	0.087	2.539	447	1.166	1.3
2003	1.601	0.496	0.158	0.174	0.064	0.088	2.580	450	1.185	1.6
2004	1.627	0.496	0.160	0.174	0.063	0.090	2.610	453	1.198	1.1
2005	1.655	0.495	0.162	0.173	0.063	0.092	2.640	455	1.212	1.2
2006	1.685	0.495	0.165	0.173	0.062	0.095	2.675	458	1.228	1.3
2007	1.710	0.498	0.167	0.172	0.062	0.097	2.706	461	1.242	1.1
2008	1.739	0.500	0.170	0.171	0.061	0.100	2.742	464	1.259	1.3
2009	1.760	0.502	0.172	0.171	0.061	0.103	2.768	466	1.271	1.0
2010	1.791	0.504	0.174	0.170	0.060	0.106	2.805	470	1.288	1.3
2011	1.818	0.505	0.178	0.169	0.060	0.109	2.838	473	1.303	1.2
2012	1.846	0.507	0.181	0.168	0.059	0.111	2.873	476	1.319	1.2
2013	1.884	0.509	0.184	0.167	0.059	0.114	2.916	479	1.338	1.5
2014	1.909	0.500	0.187	0.166	0.058	0.116	2.936	481	1.356	1.6

The table shows that CO_2 dominates the total forcing, with methane and chlorofluorocarbons (CFC) becoming relatively smaller contributors to the total forcing over time. The five major greenhouse gases account for about 96% of the direct radiative forcing by long-lived greenhouse gas increases since 1750. The remaining 4% is contributed by the 15 minor halogenated gases.

The table also includes an "Annual Greenhouse Gas Index" (AGGI), which is defined as the ratio of the total direct radiative forcing due to long-lived greenhouse gases for any year for which adequate global measurements exist to that which was present in 1990. 1990 was chosen because it is the baseline year for the Kyoto Protocol. This index is a measure of the inter-annual changes in conditions that affect carbon dioxide emission and uptake, methane and nitrous oxide sources and sinks, the decline in the atmospheric abundance of ozone-depleting chemicals related to the Montreal Protocol. and the increase in their substitutes (hydrogenated CFCs (HCFCs) and hydrofluorocarbons (HFC). Most of this increase is related to CO_2. For 2013, the AGGI was 1.34 (representing an increase in total direct radiative forcing of 34% since 1990). The increase in CO_2 forcing alone since 1990 was about 46%. The decline in CFCs considerably tempered the increase in net radiative forcing.

Deforestation and Climate Change

Deforestation is one of the main causes of climate change. It is the second largest anthropogenic source of carbon dioxide to the atmosphere, after fossil fuel combustion. Deforestation and forest degradation contribute to atmospheric greenhouse gas emissions through combustion of forest biomass and decomposition of remaining plant material and soil carbon. It used to account for more than 20% of carbon dioxide emissions, but it's currently somewhere around the 10% mark. By 2008, deforestation was 12% of total CO_2, or 15% if peatlands are included. These proportions are likely to have fallen since given the continued rise of fossil fuel use.

Averaged over all land and ocean surfaces, temperatures warmed roughly 1.53 °F (0.85 °C) between 1880 and 2012, according to the Intergovernmental Panel on Climate Change. In the Northern Hemisphere, 1983 to 2012 were the warmest 30-year period of the last 1400 years.

Effect on Climate Change

Decrease in Biodiversity

A 2007 study conducted by the National Science Foundation found that biodiversity and genetic diversity are codependent—that diversity among species requires diversity within a species, and vice versa. "If any one type is removed from the system, the cycle can break down, and the community becomes dominated by a single species."

Counteracting Climate Change

Reforestation

Reforestation is the natural or intentional restocking of existing forests and woodlands that have been depleted, usually through deforestation. It is the reestablishment of forest cover either naturally or artificially. Similar to the other methods of forestation, reforestation can be very effective because a single tree can absorb as much as 48 pounds of carbon dioxide per year and can sequester 1 ton of carbon dioxide by the time it reaches 40 years old.

Afforestation

Afforestation is the establishment of a forest or stand of trees in an area where there was no forest.

China

Although China has set official goals for reforestation, these goals were set for an 80-year time horizon and were not significantly met by 2008. China is trying to correct these problems with projects such as the Green Wall of China, which aims to replant forests and halt the expansion of the Gobi Desert. A law promulgated in 1981 requires that every school student over the age of 11 plant at least one tree per year. But average success rates, especially in state-sponsored plantings, remains relatively low. And even the properly planted trees have had great difficulty surviving the combined impacts of prolonged droughts, pest infestation and fires. Nonetheless, China currently has the highest afforestation rate of any country or region in the world, with 4.77 million hectares (47,000 square kilometers) of afforestation in 2008.

Japan

The primary goal of afforestation projects in Japan is to develop the forest structure of the nation and to maintain the biodiversity found in the Japanese wilderness. The Japanese temperate rainforest is scattered throughout the Japanese archipelago and is home to many endemic species that are not naturally found anywhere else. As development of the country's caused a decline in forest cover, a reduction in biodiversity was seen in those areas.

Agroforestry

Agroforestry or agro-sylviculture is a land use management system in which trees or shrubs are grown around or among crops or pastureland. It combines agricultural and forestry technologies to create more diverse, productive, profitable, healthy, and sustainable land-use systems.

Projects and Foundations

Arbor Day Foundation

Founded in 1972, the centennial of the first Arbor Day observance in the 19th century, the Foundation has grown to become the largest nonprofit membership organization dedicated to planting trees, with over one million members, supporters, and valued partners. They work on projects focused on planting trees around campuses, low-income communities, and communities that have been affected by natural disasters among other places.

Billion Tree Campaign

The Billion Tree Campaign was launched in 2006 by the United Nations Environment Programme (UNEP) as a response to the challenges of global warming, as well as to a wider array of sustainability challenges, from water supply to biodiversity loss. Its initial target was the planting of one billion trees in 2007. Only one year later in 2008, the campaign's objective was raised to 7 billion trees—a target to be met by the climate change conference that was held in Copenhagen, Denmark in December 2009. Three months before the conference, the 7 billion planted trees mark had been surpassed. In December 2011, after more than 12 billion trees had been planted, UNEP formally handed management of the program over to the not-for-profit Plant-for-the-Planet initiative, based in Munich, Germany.

The Amazon Fund (Brazil)

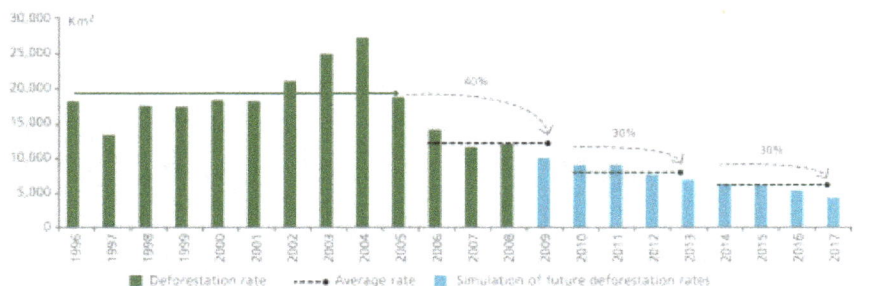

Four-year plan to reduce in deforestation in the Amazon

Considered the largest reserve of biological diversity in the world, the Amazon Basin is also the largest Brazilian biome, taking up almost half the nation's territory. The Amazon Basin corresponds to two fifths of South America's territory. Its area of approximately seven million square kilometers covers the largest hydrographic network on the planet, through which runs about one fifth of the fresh water on the world's surface. Deforestation in the Amazon rainforest is a major cause to climate change due to the decreasing number of trees available to capture increasing carbon dioxide levels in the atmosphere.

The Amazon Fund is aimed at raising donations for non-reimbursable investments in efforts to prevent, monitor and combat deforestation, as well as to promote the preservation and sustainable use of forests in the Amazon Biome, under the terms of Decree N.º 6,527, dated August 1, 2008. The Amazon Fund supports the following areas: management of public forests and protected areas, environmental control, monitoring and inspection, sustainable forest management, economic activities created with sustainable use of forests, ecological and economic zoning, territorial arrangement and agricultural regulation, preservation and sustainable use of biodiversity, and recovery of deforested areas. Besides those, the Amazon Fund may use up to 20% of its donations to support the development of systems to monitor and control deforestation in other Brazilian biomes and in biomes of other tropical countries.

Global Dimming

Global dimming is the gradual reduction in the amount of global direct irradiance at the Earth's surface that was observed for several decades after the start of systematic measurements in the 1950s. The effect varies by location, but worldwide it has been estimated to be of the order of a 4% reduction over the three decades from 1960–1990. However, after discounting an anomaly caused by the eruption of Mount Pinatubo in 1991, a very slight reversal in the overall trend has been observed.

Dozens of fires burning on the surface (red dots) and a thick pall of smoke and haze (greyish pixels) filling the skies overhead in Eastern China. Smoke, pollution and other air particles are linked to global dimming. Photo taken by MODIS aboard NASA's Aqua satellite.

Global dimming is thought to have been caused by an increase in particulates such as sulfate aerosols in the atmosphere due to human action.

It has interfered with the hydrological cycle by reducing evaporation and may have reduced rainfall in some areas. Global dimming also creates a cooling effect that may have partially counteracted the effect of greenhouse gases on global warming.

Causes and Effects

It is thought that global dimming is probably due to the increased presence of aerosol particles in the atmosphere caused by human action. Aerosols and other particulates absorb solar energy and reflect sunlight back into space. The pollutants can also become nuclei for cloud droplets. Water droplets in clouds coalesce around the particles. Increased pollution causes more particulates and thereby creates clouds consisting of a greater number of smaller droplets (that is, the same amount of water is spread over more droplets). The smaller droplets make clouds more reflective, so that more incoming sunlight is reflected back into space and less reaches the Earth's surface. This same effect also reflects radiation from below, trapping it in the lower atmosphere. In models, these smaller droplets also decrease rainfall.

Clouds intercept both heat from the sun and heat radiated from the Earth. Their effects are complex and vary in time, location, and altitude. Usually during the daytime the interception of sunlight predominates, giving a cooling effect; however, at night the re-radiation of heat to the Earth slows the Earth's heat loss, this causes storms and subsequent flood rains and flooding.

Research

In the late-1960s, Mikhail Ivanovich Budyko worked with simple two-dimensional energy-balance climate models to investigate the reflectivity of ice. He found that the ice-albedo feedback created a positive feedback loop in the Earth's climate system. The more snow and ice, the more solar radiation is reflected back into space and hence the colder Earth grows and the more it snows. Other studies found that pollution or a volcano eruption could provoke the onset of an ice age.

In the mid-1980s, Atsumu Ohmura, a geography researcher at the Swiss Federal Institute of Technology, found that solar radiation striking the Earth's surface had declined by more than 10% over the three previous decades. His findings appeared to contradict global warming — the global temperature had been generally rising since the 70s. Less light reaching the earth seemed to mean that it should cool. Ohmura published his findings "Secular variation of global radiation in Europe" in 1989. This was soon followed by others: Viivi Russak in 1990 "Trends of solar radiation, cloudiness and atmospheric transparency during recent decades in Estonia", and Beate Liepert in 1994 "Solar radiation in Germany — Observed trends and an assessment of their causes". Dimming has also been observed in sites all over the former Soviet Union. Gerry Stanhill who studied these declines worldwide in many papers coined the term "global dimming".

Independent research in Israel and the Netherlands in the late 1980s showed an apparent reduction in the amount of sunlight, despite widespread evidence that the climate was becoming hotter. The rate of dimming varies around the world but is on average estimated at around 2–3% per decade. The trend reversed in the early 1990s. It is difficult to make a precise measurement, due to the difficulty in accurately calibrating the instruments used, and the problem of spatial coverage. Nonetheless, the effect is almost certainly present.

The effect (2–3%, as above) is due to changes within the Earth's atmosphere; the value of the solar radiation at the top of the atmosphere has not changed by more than a fraction of this amount.

Smog, seen here at the Golden Gate Bridge, is a likely contributor to global dimming.

The effect varies greatly over the planet, but estimates of the terrestrial surface average value are:

- 5.3% (9 W/m²); over 1958–85 (Stanhill and Moreshet, 1992)

- 2%/decade over 1964–93 (Gilgen *et al.*, 1998)

- 2.7%/decade (total 20 W/m²); up to 2000 (Stanhill and Cohen, 2001)

- 4% over 1961–90 (Liepert 2002)

Note that these numbers are for the terrestrial surface and not really a global average. Whether dimming (or brightening) occurred over the ocean has been a bit of an unknown though a specific measurement measured effects some 400 miles (643.7 km) from India over the Indian Ocean towards the Maldives Islands. Regional effects probably dominate but are not strictly confined to the land area, and the effects will be driven by regional air circulation. A 2009 review by Wild et al. found that widespread variation in regional and time effects. There was solar brightening beyond 2000 at numerous stations in Europe, the United States, and Korea. The brightening seen at sites in Antarctica during the 1990s, influenced by recovering from the Mount Pinatubo volcanic eruption in 1991, fades after 2000. The brightening tendency also seems to level off at sites in Japan. In China there is some indication for a renewed dimming, after the stabilization in the 1990s. A continuation of the long-lasting dimming is also noted at the sites in India. Overall, the available data suggest continuation of the brightening beyond the year 2000 at numerous locations, yet less pronounced and coherent than during the 1990s, with more regions with no clear changes or declines. Therefore, globally, greenhouse warming after 2000 may be less modulated by surface solar variations than in prior decades. The largest reductions are found in the northern hemisphere mid-latitudes. Visible light and infrared radiation seem to be most affected rather than the ultraviolet part of the spectrum.

Pan Evaporation Data

Over the last 50 or so years, pan evaporation has been carefully monitored. For decades, nobody took much notice of the pan evaporation measurements. But in the 1990s in Europe, Israel, and

North America, scientists spotted something that at the time was considered very strange: the rate of evaporation was falling although they had expected it to increase due to global warming. The same trend has been observed in China over a similar period. A decrease in solar irradiance is cited as the driving force. However, unlike in other areas of the world, in China the decrease in solar irradiance was not always accompanied by an increase in cloud cover and precipitation. It is believed that aerosols may play a critical role in the decrease of solar irradiance in China.

BBC Horizon producer David Sington believes that many climate scientists regard the pan-evaporation data as the most convincing evidence of solar dimming. Pan evaporation experiments are easy to reproduce with low-cost equipment. There are many pans used for agriculture all over the world and in many instances the data have been collected for nearly a half century. However, pan evaporation depends on factors besides net radiation from the sun. The other two major factors are vapor pressure deficit and wind speed. The ambient temperature turns out to be a negligible factor. The pan evaporation data corroborates the data gathered by radiometer and fills in the gaps in the data obtained using pyranometers. With adjustments to these factors, pan evaporation data has been compared to results of climate simulations.

Probable Causes

NASA photograph showing aircraft contrails and natural clouds. The temporary disappearance of contrails over North America due to plane groundings after the September 11, 2001 attacks, and the resulting increase in diurnal temperature range gave empirical evidence of the effect of thin ice clouds at the Earth's surface.

The incomplete combustion of fossil fuels (such as diesel) and wood releases black carbon into the air. Though black carbon, most of which is soot, is an extremely small component of air pollution at land surface levels, the phenomenon has a significant heating effect on the atmosphere at altitudes above two kilometers (6,562 ft). Also, it dims the surface of the ocean by absorbing solar radiation.

Experiments in the Maldives (comparing the atmosphere over the northern and southern islands) in the 1990s showed that the effect of macroscopic pollutants in the atmosphere at that time (blown south from India) caused about a 10% reduction in sunlight reaching the surface in the area under the pollution cloud — a much greater reduction than expected from the presence of the particles themselves. Prior to the research being undertaken, predictions were of a 0.5–1% effect from particulate matter; the variation from prediction may be explained by cloud formation with the particles acting as the focus for droplet creation. Clouds are very effective at reflecting light back out into space.

The phenomenon underlying global dimming may also have regional effects. While most of the earth has warmed, the regions that are downwind from major sources of air pollution (specifically sulfur dioxide emissions) have generally cooled. This may explain the cooling of the eastern United States relative to the warming western part.

However some research shows that black carbon will increase global warming, being second only to CO_2. They believe that soot will absorb solar energy and transport it to other areas such as the Himalayas where glacial melting occurs. It can also darken Arctic ice reducing reflectivity and increasing absorption of solar radiation.

Some climate scientists have theorized that aircraft contrails (also called vapor trails) are implicated in global dimming, but the constant flow of air traffic previously meant that this could not be tested. The near-total shutdown of civil air traffic during the three days following the September 11, 2001 attacks afforded a unique opportunity in which to observe the climate of the United States absent from the effect of contrails. During this period, an increase in diurnal temperature variation of over 1 °C (1.8 °F) was observed in some parts of the U.S., i.e. aircraft contrails may have been raising nighttime temperatures and/or lowering daytime temperatures by much more than previously thought.

Airborne volcanic ash can reflect the Sun's rays back into space and thereby contribute to cooling the planet. Dips in earth temperatures have been observed after large volcano eruptions such as Mount Agung in Bali that erupted in 1963, El Chichon (Mexico) 1983, Ruiz (Colombia) 1985, and Pinatubo (Philippines) 1991. But even for major eruptions, the ash clouds remain only for relatively short periods.

Recent reversal of the Trend

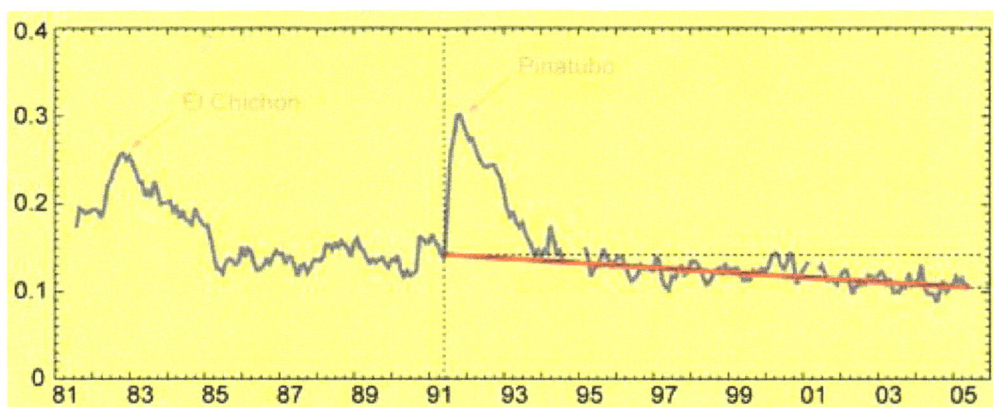

Sun-blocking aerosols around the world steadily declined (red line) since the 1991 eruption of Mount Pinatubo, according to satellite estimates. Credit: Michael Mishchenko, NASA

Wild *et al.*, using measurements over land, report brightening since 1990, and Pinker *et al.* found that slight dimming continued over land while brightening occurred over the ocean. Hence, over the land surface, Wild *et al.* and Pinker *et al.* disagree. A 2007 NASA sponsored satellite-based study sheds light on the puzzling observations by other scientists that the amount of sunlight reaching Earth's surface had been steadily declining in recent decades, began to reverse around 1990. This switch from a "global dimming" trend to a "brightening" trend happened just as global aerosol levels started to decline.

It is likely that at least some of this change, particularly over Europe, is due to decreases in airborne pollution. Most governments of developed nations have taken steps to reduce aerosols released into the atmosphere, which helps reduce global dimming.

Sulfate aerosols have declined significantly since 1970 with the Clean Air Act in the United States and similar policies in Europe. The Clean Air Act was strengthened in 1977 and 1990. According to the EPA, from 1970 to 2005, total emissions of the six principal air pollutants, including PM's, dropped by 53% in the US. In 1975, the masked effects of trapped greenhouse gases finally started to emerge and have dominated ever since.

The Baseline Surface Radiation Network (BSRN) has been collecting surface measurements. BSRN was started in the early 1990s and updated the archives in this time. Analysis of recent data reveals that the surface of the planet has brightened by about 4% in the past decade. The brightening trend is corroborated by other data, including satellite analyses.

Relationship to Hydrological Cycle

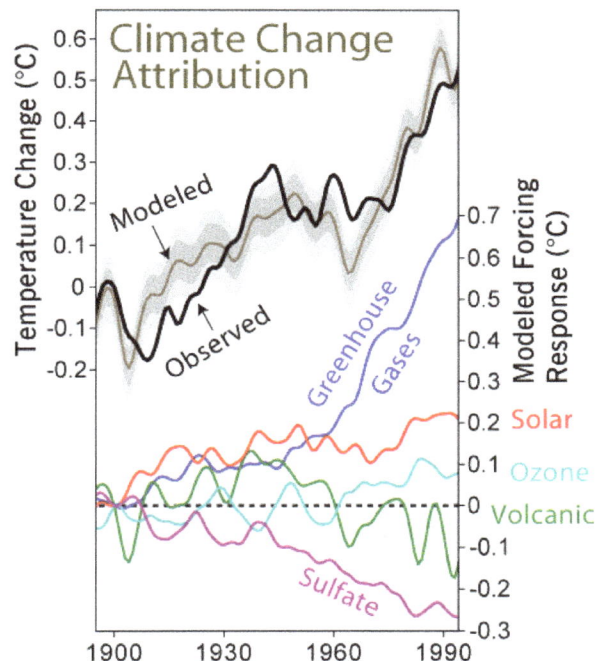

This figure shows the level of agreement between a climate model driven by five factors and the historical temperature record. The negative component identified as "sulfate" is associated with the aerosol emissions blamed for global dimming.

Pollution produced by humans may be seriously weakening the Earth's water cycle — reducing rainfall and threatening fresh water supplies. A 2001 study by researchers at the Scripps Institution of Oceanography suggests that tiny particles of soot and other pollutants have a significant effect on the hydrological cycle. According to Veerabhadran Ramanathan, "the energy for the hydrological cycle comes from sunlight. As sunlight heats the ocean, water escapes into the atmosphere and falls out as rain. So as aerosols cut down sunlight by large amounts, they may be spinning down the hydrological cycle of the planet."

Large scale changes in weather patterns may also have been caused by global dimming. Climate modelers speculatively suggest that this reduction in solar radiation at the surface may have led to the failure of the monsoon in sub-Saharan Africa during the 1970s and 1980s, together with the associated famines such as the Sahel drought, caused by Northern hemisphere pollution cooling the Atlantic. Because of this, the Tropical rain belt may not have risen to its northern latitudes, thus causing an absence of seasonal rains. This claim is not universally accepted and is very difficult to test. However a 2009 Chinese study of 50 years of continuous data found that, though most parts of eastern China saw no significant change in the amount of water held by the atmosphere, light rains had decreased. The researchers then modeled the effect of aerosols and also concluded the overall effect was that water drops in polluted cases are up to 50 percent smaller than in pristine skies. They concluded smaller size impedes the formation of rain clouds, and the falling of light rain is beneficial for agriculture. This was a different effect than reducing solar irradiance, but still a direct result from the presence of aerosols.

The 2001 study by researchers at the Scripps Institution of Oceanography concluded that the imbalance between global dimming and global warming at the surface leads to weaker turbulent heat fluxes to the atmosphere. This means globally reduced evaporation and hence precipitation occur in a dimmer and warmer world, which could ultimately lead to a more humid atmosphere in which it rains less.

A natural form of large scale environmental shading/dimming has been identified that affected the 2006 northern hemisphere hurricane season. The NASA study found that several major dust storms in June and July in the Sahara desert sent dust drifting over the Atlantic Ocean and through several effects caused cooling of the waters, thus dampening the development of hurricanes.

Relationship to Global Warming

Some scientists now consider that the effects of global dimming have masked the effect of global warming to some extent and that resolving global dimming may therefore lead to increases in predictions of future temperature rise. According to Beate Liepert, "We lived in a global warming plus a global dimming world and now we are taking out global dimming. So we end up with the global warming world, which will be much worse than we thought it will be, much hotter." The magnitude of this masking effect is one of the central problems in climate change with significant implications for future climate changes and policy responses to global warming.

Interactions between the two theories for climate modification have also been studied, as global warming and global dimming are neither mutually exclusive nor contradictory. In a paper published on March 8, 2005 in the American Geophysical Union's Geophysical Research Letters, a research team led by Anastasia Romanou of Columbia University's Department of Applied Physics and Mathematics, New York, also showed that the apparently opposing forces of global warming and global dimming can occur at the same time. Global dimming interacts with global warming by blocking sunlight that would otherwise cause evaporation and the particulates bind to water droplets. Water vapor is the major greenhouse gas. On the other hand, global dimming is affected by evaporation and rain. Rain has the effect of clearing out polluted skies.

Brown clouds have been found to amplify global warming according to Veerabhadran Ramanathan, an atmospheric chemist at the Scripps Institution of Oceanography in La Jolla, CA. "The conventional thinking is that brown clouds have masked as much as 50 percent of global warming by greenhouse gases through so-called global dimming... While this is true globally, this study reveals that over southern and eastern Asia, the soot particles in the brown clouds are in fact amplifying the atmospheric warming trend caused by greenhouse gases by as much as 50 percent."

Possible Use to Mitigate Global Warming

Some scientists have suggested using aerosols to stave off the effects of global warming as an emergency geoengineering measure. In 1974, Mikhail Budyko suggested that if global warming became a problem, the planet could be cooled by burning sulfur in the stratosphere, which would create a haze. An increase in planetary albedo of just 0.5 percent is sufficient to halve the effect of a CO_2 doubling.

The simplest solution would be to simply emit more sulfates, which would end up in troposphere - the lowest part of the atmosphere. If this were done, Earth would still face many problems, such as:

- Using sulfates causes environmental problems such as acid rain

- Using carbon black causes human health problems

- Dimming causes ecological problems such as changes in evaporation and rainfall patterns

- Droughts and/or increased rainfall cause problems for agriculture

- Aerosol has a relatively short lifetime

The solution advocated is transporting sulfates into the next higher layer of the atmosphere - stratosphere. Aerosols in the stratosphere last years instead of weeks - so only a relatively smaller (though still large) amount of sulfate emissions would be necessary, and side effects would be less. This would require developing an efficient way to transport large amounts of gases into stratosphere, many of which have been proposed though none are known to be effective or economically viable.

In a blog post, Gavin Schmidt stated that "Ideas that we should increase aerosol emissions to counteract global warming have been described as a 'Faustian bargain' because that would imply an ever increasing amount of emissions in order to match the accumulated greenhouse gas in the atmosphere, with ever increasing monetary and health costs."

References

- Grosvenor, Edwin S. and Morgan Wesson. Alexander Graham Bell: The Life and Times of the Man Who Invented the Telephone. New York: Harry N. Abrahms, Inc., 1997, p. 274, ISBN 0-8109-4005-1.

- Brian Shmaefsky (2004). Favorite demonstrations for college science: an NSTA Press journals collection. NSTA Press. p. 57. ISBN 978-0-87355-242-4.

- Lockwood, John G. (1979). Causes of Climate. Lecture notes in mathematics 1358. New York: John Wiley & Sons. p. 162. ISBN 0-470-26657-0.

- "Study: Loss Of Genetic Diversity Threatens Species Diversity". Environmental News Network. 26 September 2007. Retrieved 27 October 2014.

- "Commit to Action - Join the Billion Tree Campaign!". UNEP. United Nations Environment Programme (UNEP). Retrieved 22 October 2014.

- "UNEP Billion Tree Campaign Hands Over to the Young People of the Plant-for-the-Planet Foundation". UNEP. Retrieved 28 October 2014.

- "How much has the Global Temperature Risen in the Last 100 Years?". National Center for Atmospheric Research. University Corporation for Atmospheric Research. Retrieved 20 October 2014.

- "Introduction to Atmospheric Chemistry, by Daniel J. Jacob, Princeton University Press, 1999. Chapter 7, "The Greenhouse Effect"". Acmg.seas.harvard.edu. Retrieved 2010-10-15.

- "Titan: Greenhouse and Anti-greenhouse :: Astrobiology Magazine - earth science - evolution distribution Origin of life universe - life beyond :: Astrobiology is study of earth". Astrobio.net. Retrieved 2010-10-15.

Integrated Study of Greenhouse Gases

The atmosphere is a mixture of different gases that facilitate chemical and physical activity. But a sharp increase or decline in any one of these gases during a short interval can cause imbalances in the atmosphere and the surface. The chapter strategically encompasses and incorporates the major components and key concepts of greenhouse gases, providing a complete understanding.

Greenhouse Gas

A greenhouse gas (sometimes abbreviated GHG) is a gas in an atmosphere that absorbs and emits radiation within the thermal infrared range. This process is the fundamental cause of the greenhouse effect. The primary greenhouse gases in Earth's atmosphere are water vapor, carbon dioxide, methane, nitrous oxide, and ozone. Without greenhouse gases, the average temperature of Earth's surface would be about −18 °C (0 °F), rather than the present average of 15 °C (59 °F). In the Solar System, the atmospheres of Venus, Mars and Titan also contain gases that cause a greenhouse effect.

Human activities since the beginning of the Industrial Revolution (taken as the year 1750) have produced a 40% increase in the atmospheric concentration of carbon dioxide, from 280 ppm in 1750 to 400 ppm in 2015. This increase has occurred despite the uptake of a large portion of the emissions by various natural "sinks" involved in the carbon cycle. Anthropogenic carbon dioxide (CO_2) emissions (i.e. emissions produced by human activities) come from combustion of carbon-based fuels, principally coal, oil, and natural gas, along with deforestation, soil erosion and animal agriculture.

It has been estimated that if greenhouse gas emissions continue at the present rate, Earth's surface temperature could exceed historical values as early as 2047, with potentially harmful effects on ecosystems, biodiversity and the livelihoods of people worldwide. Recent estimates suggest that on the current emissions trajectory the Earth could pass a threshold of 2°C global warming, which the United Nations' IPCC designated as the upper limit for "dangerous" global warming, by 2036.

Gases in Earth's Atmosphere

Greenhouse Gases

Greenhouse gases are those that absorb and emit infrared radiation in the wavelength range emitted by Earth. In order, the most abundant greenhouse gases in Earth's atmosphere are:

- Water vapor (H2O)
- Carbon dioxide (CO_2)

- Methane (CH4)

- Nitrous oxide (N2O)

- Ozone (O3)

- Chlorofluorocarbons (CFCs)

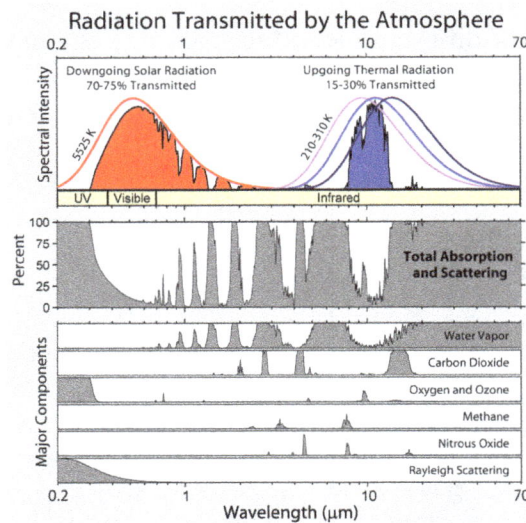

Atmospheric absorption and scattering at different wavelengths of electromagnetic waves. The largest absorption band of carbon dioxide is in the infrared.

Atmospheric concentrations of greenhouse gases are determined by the balance between sources (emissions of the gas from human activities and natural systems) and sinks (the removal of the gas from the atmosphere by conversion to a different chemical compound). The proportion of an emission remaining in the atmosphere after a specified time is the "airborne fraction" (AF). More precisely, the annual airborne fraction is the ratio of the atmospheric increase in a given year to that year's total emissions. Over the last 50 years (1956–2006) the airborne fraction for CO_2 has been increasing at $0.25 \pm 0.21\%$/year.

Non-greenhouse Gases

The major atmospheric constituents, nitrogen (N2), oxygen (O2), and argon (Ar), are not greenhouse gases. This is because molecules containing two atoms of the same element such as N 2 and O2 and monatomic molecules such as argon (Ar) have no net change in the distribution of their electrical charges when they vibrate and hence are almost totally unaffected by infrared radiation. Although molecules containing two atoms of different elements such as carbon monoxide (CO) or hydrogen chloride (HCl) absorb infrared radiation, these molecules are short-lived in the atmosphere owing to their reactivity and solubility. Therefore, they do not contribute significantly to the greenhouse effect and usually are omitted when discussing greenhouse gases.

Indirect Radiative Effects

Some gases have indirect radiative effects (whether or not they are greenhouse gases them-selves). This happens in two main ways. One way is that when they break down in the atmosphere they produce another greenhouse gas. For example, methane and carbon monoxide (CO) are oxidized to give carbon dioxide (and methane oxidation also produces water vapor; that will be considered below). Oxidation of CO to CO_2 directly produces an unambiguous increase in radiative forcing although the reason is subtle. The peak of the thermal IR emission from Earth's surface is very close to a strong vibrational absorption band of CO_2 (667 cm⁻¹). On the other hand, the single CO vibrational band only absorbs IR at much higher frequencies (2145 cm⁻¹), where the ~300 K thermal emission of the surface is at least a factor of ten lower. On the other hand, oxidation of methane to CO_2, which requires reactions with the OH radical, produces an instantaneous reduction, since CO_2 is a weaker greenhouse gas than methane; but it has a longer lifetime. As described below this is not the whole story, since the oxidations of CO and CH 4 are intertwined by both consuming OH radicals. In any case, the calculation of the total radiative effect needs to include both the direct and indirect forcing.

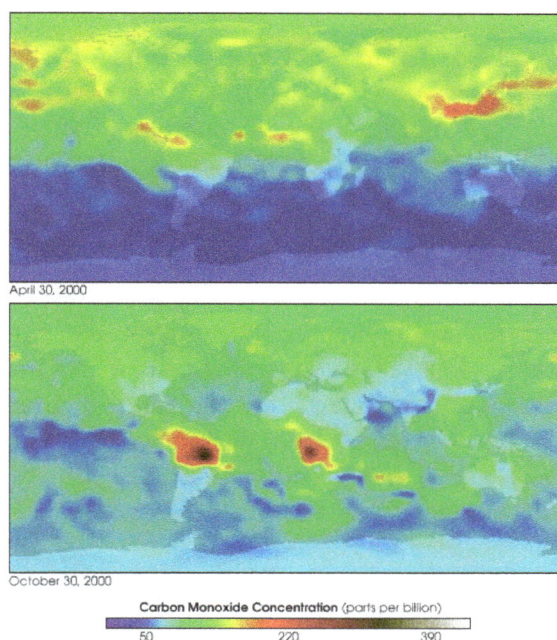

The false colors in this image represent concentrations of carbon monoxide in the lower atmosphere, ranging from about 390 parts per billion (dark brown pixels), to 220 parts per billion (red pixels), to 50 parts per billion (blue pixels).

A second type of indirect effect happens when chemical reactions in the atmosphere involving these gases change the concentrations of greenhouse gases. For example, the destruction of non-methane volatile organic compounds (NMVOCs) in the atmosphere can produce ozone. The size of the indirect effect can depend strongly on where and when the gas is emitted.

Methane has a number of indirect effects in addition to forming CO_2. Firstly, the main chemical that destroys methane in the atmosphere is the hydroxyl radical (OH). Methane reacts with OH and so more methane means that the concentration of OH goes down. Effectively, methane increases its own atmospheric lifetime and therefore its overall radiative effect. The second effect is that the oxidation of methane can produce ozone. Thirdly, as well as making CO_2 the oxidation of methane produces water; this is a major source of water vapor in the stratosphere, which is otherwise very dry. CO and NMVOC also produce CO_2 when they are oxidized. They remove OH from the atmosphere and this leads to higher concentrations of methane. The surprising effect of this is that the global warming potential of CO is three times that of CO_2. The same process that converts NMVOC to carbon dioxide can also lead to the formation of tropospheric ozone. Halocarbons have an indirect effect because they destroy stratospheric ozone. Finally hydrogen can lead to ozone production and CH 4 increases as well as producing water vapor in the stratosphere.

Contribution of Clouds to Earth's Greenhouse Effect

The major non-gas contributor to Earth's greenhouse effect, clouds, also absorb and emit infrared radiation and thus have an effect on radiative properties of the greenhouse gases. Clouds are water droplets or ice crystals suspended in the atmosphere.

Impacts on the Overall Greenhouse Effect

Schmidt *et al.* (2010) analysed how individual components of the atmosphere contribute to the total greenhouse effect. They estimated that water vapor accounts for about 50% of Earth's greenhouse effect, with clouds contributing 25%, carbon dioxide 20%, and the minor greenhouse gases and aerosols accounting for the remaining 5%. In the study, the reference model atmosphere is for 1980 conditions. Image credit: NASA.

The contribution of each gas to the greenhouse effect is affected by the characteristics of that gas, its abundance, and any indirect effects it may cause. For example, the direct radiative effect of a mass of methane is about 72 times stronger than the same mass of carbon dioxide over a 20-year time frame but it is present in much smaller concentrations so that its total direct radiative effect is smaller, in part due to its shorter atmospheric lifetime. On the other hand, in addition to its direct radiative impact, methane has a large, indirect radiative effect because it contributes to ozone formation. Shindell *et al.* (2005) argue that the contribution to climate change from methane is at least double previous estimates as a result of this effect.

When ranked by their direct contribution to the greenhouse effect, the most important are:

Compound	Formula	Concentration in atmosphere (ppm)	Contribution (%)
Water vapor and clouds	H_2O	10−50,000[A]	36−72%
Carbon dioxide	CO_2	~400	9−26%
Methane	CH4	~1.8	4−9%
Ozone	O3	2−8[B]	3−7%
notes: [A] Water vapor strongly varies locally [B] The concentration in stratosphere. About 90% of the ozone in Earth's atmosphere is contained in the stratosphere.			

In addition to the main greenhouse gases listed above, other greenhouse gases include sulfur hexafluoride, hydrofluorocarbons and perfluorocarbons. Some greenhouse gases are not often listed. For example, nitrogen trifluoride has a high global warming.

Potential (GWP) but is only present in very small quantities.

Proportion of Direct Effects at a Given Moment

It is not possible to state that a certain gas causes an exact percentage of the greenhouse effect. This is because some of the gases absorb and emit radiation at the same frequencies as others, so that the total greenhouse effect is not simply the sum of the influence of each gas. The higher ends of the ranges quoted are for each gas alone; the lower ends account for overlaps with the other gases. In addition, some gases such as methane are known to have large indirect effects that are still being quantified.

Atmospheric Lifetime

Aside from water vapor, which has a residence time of about nine days, major greenhouse gases are well mixed and take many years to leave the atmosphere. Although it is not easy to know with precision how long it takes greenhouse gases to leave the atmosphere, there are estimates for the principal greenhouse gases. Jacob (1999) defines the lifetime τ of an atmospheric species X in a one-box model as the average time that a molecule of X remains in the box. Mathematically τ can be defined as the ratio of the mass m (in kg) of X in the box to its removal rate, which is the sum of the flow of X out of the box (F_{out}), chemical loss of X (L), and deposition of X (L) (all in kg/s): $\tau = \dfrac{m}{F_{out} + L + D}$. If one stopped pouring any of this gas into the box, then after a time τ , its concentration would be about halved.

The atmospheric lifetime of a species therefore measures the time required to restore equilibrium following a sudden increase or decrease in its concentration in the atmosphere. Individual atoms or molecules may be lost or deposited to sinks such as the soil, the oceans and other waters, or vegetation and other biological systems, reducing the excess to background concentrations. The average time taken to achieve this is the mean lifetime.

Carbon dioxide has a variable atmospheric lifetime, and cannot be specified precisely. The atmospheric lifetime of CO_2 is estimated of the order of 30–95 years. This figure accounts for CO_2 molecules being removed from the atmosphere by mixing into the ocean, photosynthesis, and other processes. However, this excludes the balancing fluxes of CO_2 into the atmosphere from the geological reservoirs, which have slower characteristic rates. Although more than half of the CO_2 emitted is removed from the atmosphere within a century, some fraction (about 20%) of emitted CO_2 remains in the atmosphere for many thousands of years. Similar issues apply to other greenhouse gases, many of which have longer mean lifetimes than CO_2. E.g., N_2O has a mean atmospheric lifetime of 114 years.

Radiative Forcing

Earth absorbs some of the radiant energy received from the sun, reflects some of it as light and reflects or radiates the rest back to space as heat. Earth's surface temperature depends on this balance between incoming and outgoing energy. If this energy balance is shifted, Earth's surface could become warmer or cooler, leading to a variety of changes in global climate.

A number of natural and man-made mechanisms can affect the global energy balance and force changes in Earth's climate. Greenhouse gases are one such mechanism. Greenhouse gases in the atmosphere absorb and re-emit some of the outgoing energy radiated from Earth's surface, causing that heat to be retained in the lower atmosphere. As explained above, some greenhouse gases remain in the atmosphere for decades or even centuries, and therefore can affect Earth's energy balance over a long time period. Factors that influence Earth's energy balance can be quantified in terms of "radiative climate forcing." Positive radiative forcing indicates warming (for example, by increasing incoming energy or decreasing the amount of energy that escapes to space), whereas negative forcing is associated with cooling.

Global Warming Potential

The global warming potential (GWP) depends on both the efficiency of the molecule as a greenhouse gas and its atmospheric lifetime. GWP is measured relative to the same **mass** of CO_2 and evaluated for a specific timescale. Thus, if a gas has a high (positive) radiative forcing but also a short lifetime, it will have a large GWP on a 20-year scale but a small one on a 100-year scale. Conversely, if a molecule has a longer atmospheric lifetime than CO_2 its GWP will increase with the timescale considered. Carbon dioxide is defined to have a GWP of 1 over all time periods.

Methane has an atmospheric lifetime of 12 ± 3 years. The 2007 IPCC report lists the GWP as 72 over a time scale of 20 years, 25 over 100 years and 7.6 over 500 years. A 2014 analysis, however, states that although methane's initial impact is about 100 times greater than that of CO_2, because of the shorter atmospheric lifetime, after six or seven decades, the impact of the two gases is about equal, and from then on methane's relative role continues to decline. The decrease in GWP at longer times is because methane is degraded to water and CO_2 through chemical reactions in the atmosphere.

Examples of the atmospheric lifetime and GWP relative to CO_2 for several greenhouse gases are given in the following table:

Atmospheric lifetime and GWP relative to CO_2 at different time horizon for various greenhouse gases.					
Gas name	Chemical formula	Lifetime (years)	Global warming potential (GWP) for given time horizon		
			20-yr	100-yr	500-yr
Carbon dioxide	CO_2	30–95	1	1	1
Methane	CH4	12	72	25	7.6
Nitrous oxide	N2O	114	289	298	153
CFC-12	CCl2F2	100	11 000	10 900	5 200
HCFC-22	CHClF2	12	5 160	1 810	549
Tetrafluoromethane	CF4	50 000	5 210	7 390	11 200
Hexafluoroethane	C2F6	10 000	8 630	12 200	18 200
Sulfur hexafluoride	SF6	3 200	16 300	22 800	32 600
Nitrogen trifluoride	NF3	740	12 300	17 200	20 700

The use of CFC-12 (except some essential uses) has been phased out due to its ozone depleting properties. The phasing-out of less active HCFC-compounds will be completed in 2030.

Carbon dioxide in Earth's atmosphere if *half* of global-warming emissions are *not* absorbed.
(NASA simulation; 9 November 2015)

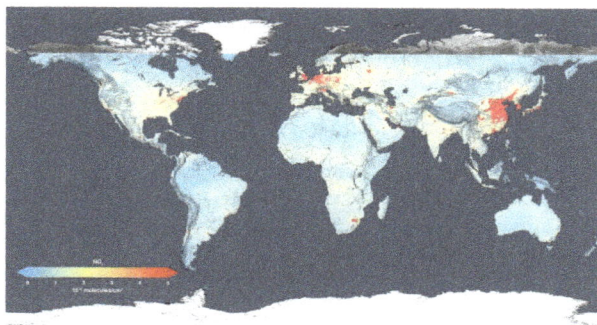

Nitrogen dioxide 2014 – global air quality levels
(released 14 December 2015).

Natural and Anthropogenic Sources

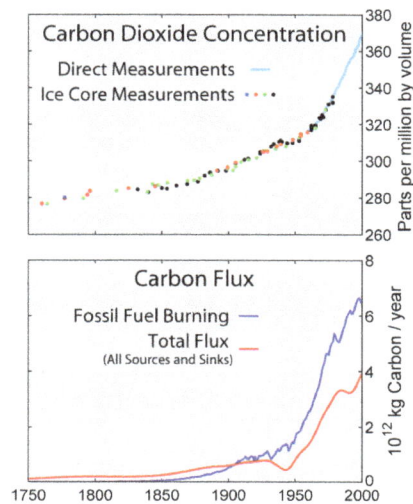

Top: Increasing atmospheric carbon dioxide levels as measured in the atmosphere and reflected in ice cores. Bottom: The amount of net carbon increase in the atmosphere, compared to carbon emissions from burning fossil fuel.

This diagram shows a simplified representation of the contemporary global carbon cycle. Changes are measured in gigatons of carbon per year (GtC/y). Canadell *et al.* (2007) estimated the growth rate of global average atmospheric CO_2 for 2000–2006 as 1.93 parts-per-million per year (4.1 petagrams of carbon per year).

Aside from purely human-produced synthetic halocarbons, most greenhouse gases have both natural and human-caused sources. During the pre-industrial Holocene, concentrations of existing gases were roughly constant. In the industrial era, human activities have added greenhouse gases to the atmosphere, mainly through the burning of fossil fuels and clearing of forests.

The 2007 Fourth Assessment Report compiled by the IPCC (AR4) noted that "changes in atmospheric concentrations of greenhouse gases and aerosols, land cover and solar radiation alter the energy balance of the climate system", and concluded that "increases in anthropogenic greenhouse gas concentrations is very likely to have caused most of the increases in global average temperatures since the mid-20th century". In AR4, "most of" is defined as more than 50%.

Abbreviations used in the two tables below: ppm = parts-per-million; ppb = parts-per-billion; ppt = parts-per-trillion; W/m² = watts per square metre

Current greenhouse gas concentrations					
Gas	Pre-1750 tropospheric concentration	Recent tropospheric concentration	Absolute increase since 1750	Percentage increase since 1750	Increased radiative forcing (W/m²)
Carbon dioxide (CO_2)	280 ppm	395.4 ppm	115.4 ppm	41.2%	1.88
Methane (CH4)	700 ppb	1893 ppb / 1762 ppb	1193 ppb / 1062 ppb	170.4% / 151.7%	0.49
Nitrous oxide (N2O)	270 ppb	326 ppb / 324 ppb	56 ppb / 54 ppb	20.7% / 20.0%	0.17
Tropospheric ozone (O3)	237 ppb	337 ppb	100 ppb	42%	0.4

Relevant to radiative forcing and/or ozone depletion; all of the following have no natural sources and hence zero amounts pre-industrial		
Gas	**Recent tropospheric concentration**	**Increased radiative forcing (W/m²)**
CFC-11 (trichlorofluoromethane) (CCl3F)	236 ppt / 234 ppt	0.061
CFC-12 (CCl2F2)	527 ppt / 527 ppt	0.169
CFC-113 (Cl2FC-CClF2)	74 ppt / 74 ppt	0.022
HCFC-22 (CHClF2)	231 ppt / 210 ppt	0.046
HCFC-141b (CH3CCl2F)	24 ppt / 21 ppt	0.0036
HCFC-142b (CH3CClF2)	23 ppt / 21 ppt	0.0042
Halon 1211 (CBrClF2)	4.1 ppt / 4.0 ppt	0.0012
Halon 1301 (CBrClF3)	3.3 ppt / 3.3 ppt	0.001
HFC-134a (CH2FCF3)	75 ppt / 64 ppt	0.0108
Carbon tetrachloride (CCl4)	85 ppt / 83 ppt	0.0143

Sulfur hexafluoride (SF6)	7.79 ppt / 7.39 ppt	0.0043
Other halocarbons	Varies by substance	collectively 0.02
Halocarbons in total		0.3574

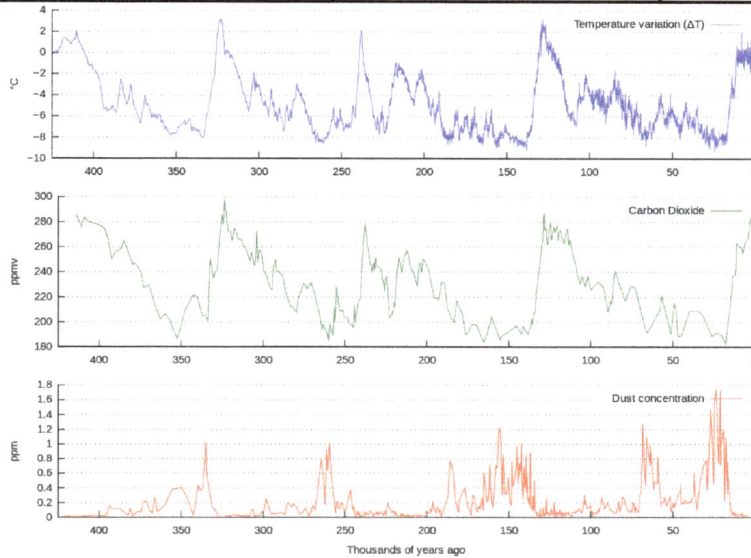

400,000 years of ice core data

Ice cores provide evidence for greenhouse gas concentration variations over the past 800,000 years. Both CO_2 and CH4 vary between glacial and interglacial phases, and concentrations of these gases correlate strongly with temperature. Direct data does not exist for periods earlier than those represented in the ice core record, a record that indicates CO_2 mole fractions stayed within a range of 180 ppm to 280 ppm throughout the last 800,000 years, until the increase of the last 250 years. However, various proxies and modeling suggests larger variations in past epochs; 500 million years ago CO_2 levels were likely 10 times higher than now. Indeed, higher CO_2 concentrations are thought to have prevailed throughout most of the Phanero-zoic eon, with concentrations four to six times current concentrations during the Mesozoic era, and ten to fifteen times current concentrations during the early Palaeozoic era until the middle of the Devonian period, about 400 Ma. The spread of land plants is thought to have reduced CO_2 concentrations during the late Devonian, and plant activities as both sources and sinks of CO_2 have since been important in providing stabilising feedbacks. Earlier still, a 200-million year period of intermittent, widespread glaciation extending close to the equator (Snowball Earth) appears to have been ended suddenly, about 550 Ma, by a colossal volcanic outgassing that raised the CO_2 concentration of the atmosphere abruptly to 12%, about 350 times modern levels, causing extreme greenhouse conditions and carbonate deposition as limestone at the rate of about 1 mm per day. This episode marked the close of the Precambrian eon, and was succeeded by the generally warmer conditions of the Phanerozoic, during which multicellular animal and plant life evolved. No volcanic carbon dioxide emission of comparable scale has occurred since. In the modern era, emissions to the atmosphere from volcanoes are only about 1% of emissions from human sources.

Ice Cores

Measurements from Antarctic ice cores show that before industrial emissions started atmospheric CO_2 mole fractions were about 280 parts per million (ppm), and stayed between 260 and 280 during the preceding ten thousand years. Carbon dioxide mole fractions in the atmosphere have gone up by approximately 35 percent since the 1900s, rising from 280 parts per million by volume to 387 parts per million in 2009. One study using evidence from stomata of fossilized leaves suggests greater variability, with carbon dioxide mole fractions above 300 ppm during the period seven to ten thousand years ago, though others have argued that these findings more likely reflect calibration or contamination problems rather than actual CO_2 variability. Because of the way air is trapped in ice (pores in the ice close off slowly to form bubbles deep within the firn) and the time period represented in each ice sample analyzed, these figures represent averages of atmospheric concentrations of up to a few centuries rather than annual or decadal levels.

Changes Since the Industrial Revolution

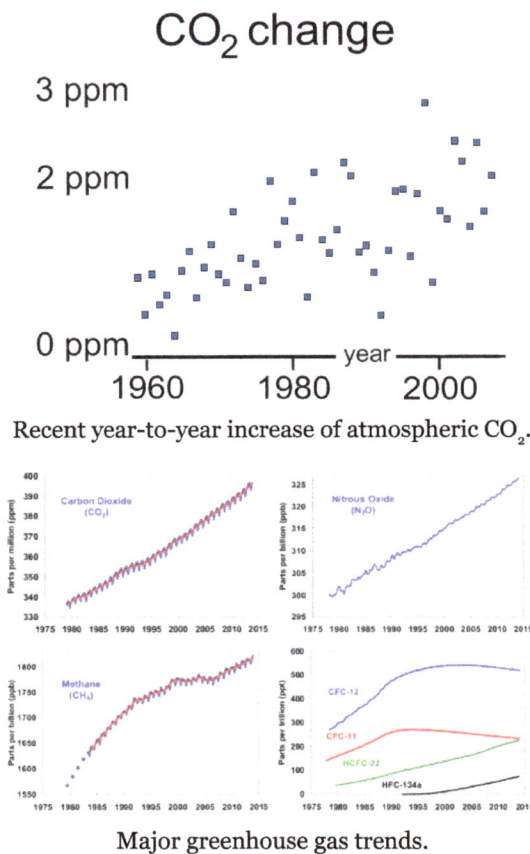

Recent year-to-year increase of atmospheric CO_2.

Major greenhouse gas trends.

Since the beginning of the Industrial Revolution, the concentrations of most of the greenhouse gases have increased. For example, the mole fraction of carbon dioxide has increased from 280 ppm by about 36% to 380 ppm, or 100 ppm over modern pre-industrial levels. The first 50 ppm increase took place in about 200 years, from the start of the Industrial Revolution to around 1973.; however the next 50 ppm increase took place in about 33 years, from 1973 to 2006.

Recent data also shows that the concentration is increasing at a higher rate. In the 1960s, the average annual increase was only 37% of what it was in 2000 through 2007.

Today, the stock of carbon in the atmosphere increases by more than 3 million tonnes per annum (0.04%) compared with the existing stock. This increase is the result of human activities by burning fossil fuels, deforestation and forest degradation in tropical and boreal regions.

The other greenhouse gases produced from human activity show similar increases in both amount and rate of increase. Many observations are available online in a variety of Atmospheric Chemistry Observational Databases.

Anthropogenic Greenhouse Gases

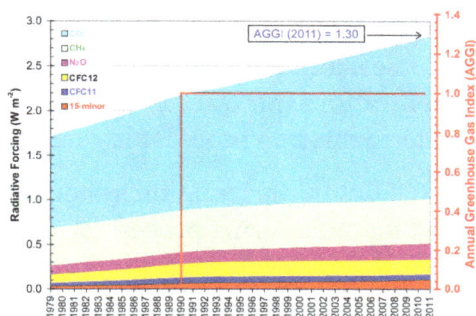

This graph shows changes in the annual greenhouse gas index (AGGI) between 1979 and 2011. The AGGI measures the levels of greenhouse gases in the atmosphere based on their ability to cause changes in Earth's climate.

This bar graph shows global greenhouse gas emissions by sector from 1990 to 2005, measured in carbon dioxide equivalents.

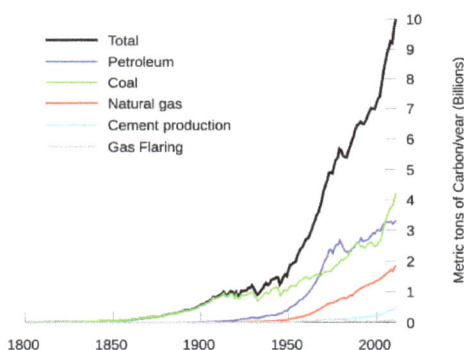

Modern global CO2 emissions from the burning of fossil fuels.

Since about 1750 human activity has increased the concentration of carbon dioxide and other greenhouse gases. Measured atmospheric concentrations of carbon dioxide are currently 100 ppm higher than pre-industrial levels. Natural sources of carbon dioxide are more than 20 times greater than sources due to human activity, but over periods longer than a few years natural sources are closely balanced by natural sinks, mainly photosynthesis of carbon compounds by plants and marine plankton. As a result of this balance, the atmospheric mole fraction of carbon dioxide remained between 260 and 280 parts per million for the 10,000 years between the end of the last glacial maximum and the start of the industrial era.

It is likely that anthropogenic (i.e., human-induced) warming, such as that due to elevated greenhouse gas levels, has had a discernible influence on many physical and biological systems. Future warming is projected to have a range of impacts, including sea level rise, increased frequencies and severities of some extreme weather events, loss of biodiversity, and regional changes in agricultural productivity.

The main sources of greenhouse gases due to human activity are:

- burning of fossil fuels and deforestation leading to higher carbon dioxide concentrations in the air. Land use change (mainly deforestation in the tropics) account for up to one third of total anthropogenic CO_2 emissions.

- livestock enteric fermentation and manure management, paddy rice farming, land use and wetland changes, pipeline losses, and covered vented landfill emissions leading to higher methane atmospheric concentrations. Many of the newer style fully vented septic systems that enhance and target the fermentation process also are sources of atmospheric methane.

- use of chlorofluorocarbons (CFCs) in refrigeration systems, and use of CFCs and halons in fire suppression systems and manufacturing processes.

- agricultural activities, including the use of fertilizers, that lead to higher nitrous oxide (N 2O) concentrations.

The seven sources of CO_2 from fossil fuel combustion are (with percentage contributions for 2000–2004):

Seven main fossil fuel combustion sources	Contribution (%)
Liquid fuels (e.g., gasoline, fuel oil)	36%
Solid fuels (e.g., coal)	35%
Gaseous fuels (e.g., natural gas)	20%
Cement production	3 %
Flaring gas industrially and at wells	< 1%
Non-fuel hydrocarbons	< 1%
"International bunker fuels" of transport not included in national inventories	4 %

Carbon dioxide, methane, nitrous oxide (N2O) and three groups of fluorinated gases (sulfur hexafluoride (SF6), hydrofluorocarbons (HFCs), and perfluorocarbons (PFCs)) are the major anthropogenic greenhouse gases and are regulated under the Kyoto Protocol international treaty, which came into force in 2005. Emissions limitations specified in the Kyoto Protocol expired in 2012. The Cancún agreement, agreed in 2010, includes voluntary pledges made by 76 countries to control emissions. At the time of the agreement, these 76 countries were collectively responsible for 85% of annual global emissions.

Although CFCs are greenhouse gases, they are regulated by the Montreal Protocol, which was motivated by CFCs' contribution to ozone depletion rather than by their contribution to global warming. Note that ozone depletion has only a minor role in greenhouse warming though the two processes often are confused in the media.

Tourism

According to UNEP global tourism is closely linked to climate change. Tourism is a significant contributor to the increasing concentrations of greenhouse gases in the atmosphere. Tourism accounts for about 50% of traffic movements. Rapidly expanding air traffic contributes about 2.5% of the production of CO_2. The number of international travelers is expected to increase from 594 million in 1996 to 1.6 billion by 2020, adding greatly to the problem unless steps are taken to reduce emissions.

Road Haulage

The road haulage industry plays a part in production of CO_2, contributing around 20% of the UK's total carbon emissions a year, with only the energy industry having a larger impact at around 39%. Average carbon emissions within the haulage industry are falling—in the thirty-year period from 1977–2007, the carbon emissions associated with a 200-mile journey fell by 21 percent; NOx emissions are also down 87 percent, whereas journey times have fallen by around a third. Due to their size, HGVs often receive criticism regarding their CO2 emissions; however, rapid development in engine technology and fuel management is having a largely positive effect.

Role of Water Vapor

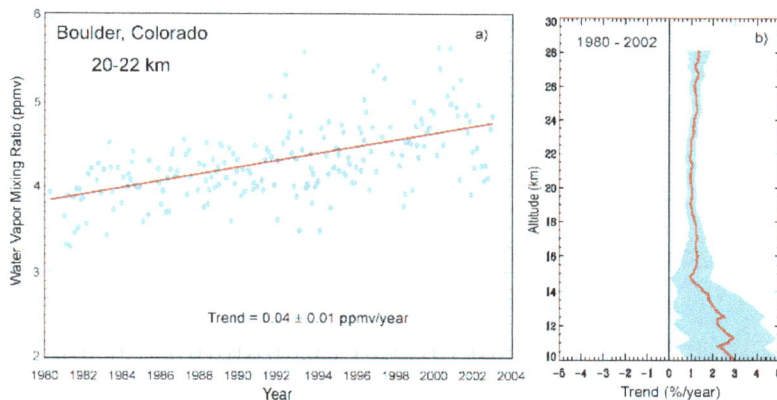

Increasing water vapor in the stratosphere at Boulder, Colorado.

Water vapor accounts for the largest percentage of the greenhouse effect, between 36% and 66% for clear sky conditions and between 66% and 85% when including clouds. Water vapor concentrations fluctuate regionally, but human activity does not significantly affect water vapor concentrations except at local scales, such as near irrigated fields. The atmospheric concentration of vapor is highly variable and depends largely on temperature, from less than 0.01% in extremely cold regions up to 3% by mass in saturated air at about 32 °C.

The average residence time of a water molecule in the atmosphere is only about nine days, compared to years or centuries for other greenhouse gases such as CH4 and CO_2. Thus, water vapor responds to and amplifies effects of the other greenhouse gases. The Clausius–Clapeyron relation establishes that more water vapor will be present per unit volume at elevated temperatures. This and other basic principles indicate that warming associated with increased concentrations of the other greenhouse gases also will increase the concentration of water vapor (assuming that the relative humidity remains approximately constant; modeling and observational studies find that this is indeed so). Because water vapor is a greenhouse gas, this results in further warming and so is a "positive feedback" that amplifies the original warming. Eventually other earth processes offset these positive feedbacks, stabilizing the global temperature at a new equilibrium and preventing the loss of Earth's water through a Venus-like runaway greenhouse effect.

Direct Greenhouse Gas Emissions

Between the period 1970 to 2004, GHG emissions (measured in CO_2-equivalent) increased at an average rate of 1.6% per year, with CO_2 emissions from the use of fossil fuels growing at a rate of 1.9% per year. Total anthropogenic emissions at the end of 2009 were estimated at 49.5 gigatonnes CO_2-equivalent. These emissions include CO_2 from fossil fuel use and from land use, as well as emissions of methane, nitrous oxide and other GHGs covered by the Kyoto Protocol.

At present, the primary source of CO_2 emissions is the burning of coal, natural gas, and petroleum for electricity and heat.

Regional and National Attribution of Emissions

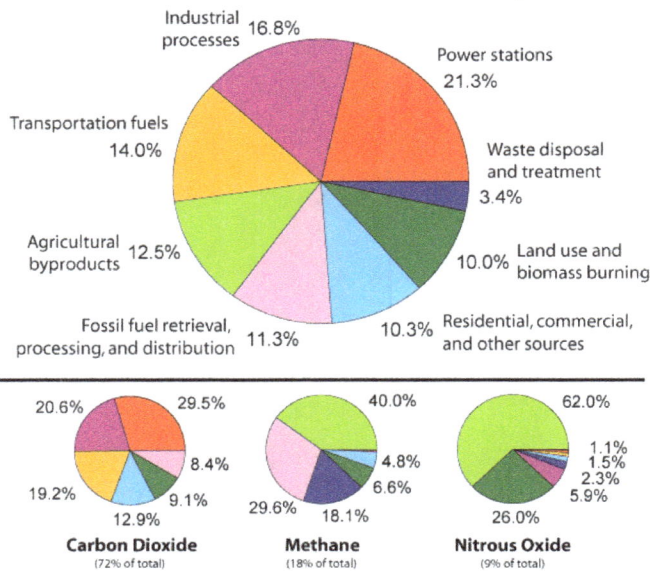

Annual Greenhouse Gas Emissions by Sector

This figure shows the relative fraction of anthropogenic greenhouse gases coming from each of eight categories of sources, as estimated by the Emission Database for Global Atmospheric Research version 3.2, fast track 2000 project . These values are intended to provide a snapshot of global annual greenhouse gas emissions in the year 2000. The top panel shows the sum over all anthropogenic greenhouse gases, weighted by their global warming potential over the next 100 years. This consists of 72% carbon dioxide, 18% methane, 8% nitrous oxide and 1% other gases. Lower panels show the comparable information for each of these three primary greenhouse gases, with the same coloring of sectors as used in the top chart. Segments with less than 1% fraction are not labeled.

There are several different ways of measuring GHG emissions, for example, World Bank (2010) for section of national emissions data. Some variables that have been reported include:

- Definition of measurement boundaries: Emissions can be attributed geographically, to the area where they were emitted (the territory principle) or by the activity principle to the territory produced the emissions. These two principles result in different totals when measuring, for example, electricity importation from one country to another, or emissions at an international airport.

- Time horizon of different GHGs: Contribution of a given GHG is reported as a CO_2 equivalent. The calculation to determine this takes into account how long that gas remains in the atmosphere. This is not always known accurately and calculations must be regularly updated to reflect new information.

- What sectors are included in the calculation (e.g., energy industries, industrial processes, agriculture etc.): There is often a conflict between transparency and availability of data.

- The measurement protocol itself: This may be via direct measurement or estimation. The four main methods are the emission factor-based method, mass balance method, predictive emissions monitoring systems, and continuous emissions monitoring systems. These methods differ in accuracy, cost, and usability.

These different measures are sometimes used by different countries to assert various policy/ethical positions on climate change (Banuri *et al.*, 1996, p. 94). This use of different measures leads to a lack of comparability, which is problematic when monitoring progress towards targets. There are arguments for the adoption of a common measurement tool, or at least the development of communication between different tools.

Emissions may be measured over long time periods. This measurement type is called historical or cumulative emissions. Cumulative emissions give some indication of who is responsible for the build-up in the atmospheric concentration of GHGs (IEA, 2007, p. 199).

The national accounts balance would be positively related to carbon emissions. The national accounts balance shows the difference between exports and imports. For many richer nations, such as the United States, the accounts balance is negative because more goods are imported than they are exported. This is mostly due to the fact that it is cheaper to produce goods outside of developed countries, leading the economies of developed countries to become increasingly dependent on services and not goods. We believed that a positive accounts balance would means that more production was occurring in a country, so more factories working would increase carbon emission levels.(Holtz-Eakin, 1995, pp.;85;101).

Emissions may also be measured across shorter time periods. Emissions changes may, for example, be measured against a base year of 1990. 1990 was used in the United Nations Framework Convention on Climate Change (UNFCCC) as the base year for emissions, and is also used in the Kyoto Protocol (some gases are also measured from the year 1995). A country's emissions may also be reported as a proportion of global emissions for a particular year.

Another measurement is of per capita emissions. This divides a country's total annual emissions by its mid-year population. Per capita emissions may be based on historical or annual emissions (Banuri *et al.*, 1996, pp. 106–107).

Land-Use Change

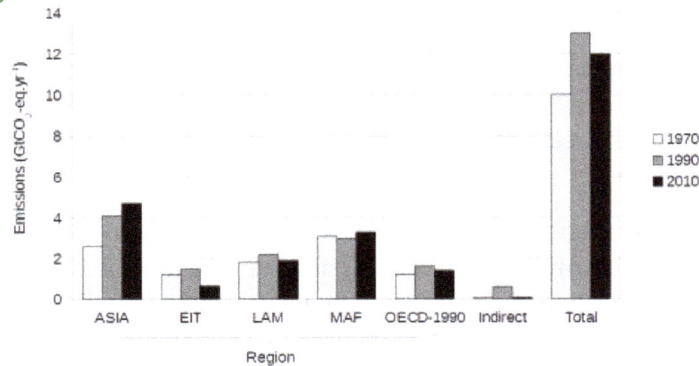

Greenhouse gas emissions from agriculture, forestry and other land use, 1970–2010.

Land-use change, e.g., the clearing of forests for agricultural use, can affect the concentration of GHGs in the atmosphere by altering how much carbon flows out of the atmosphere into carbon sinks. Accounting for land-use change can be understood as an attempt to measure "net" emissions, i.e., gross emissions from all GHG sources minus the removal of emissions from the atmosphere by carbon sinks (Banuri *et al.*, 1996, pp. 92–93).

There are substantial uncertainties in the measurement of net carbon emissions. Additionally, there is controversy over how carbon sinks should be allocated between different regions and over time (Banuri *et al.*, 1996, p. 93). For instance, concentrating on more recent changes in carbon sinks is likely to favour those regions that have deforested earlier, e.g., Europe.

Greenhouse Gas Intensity

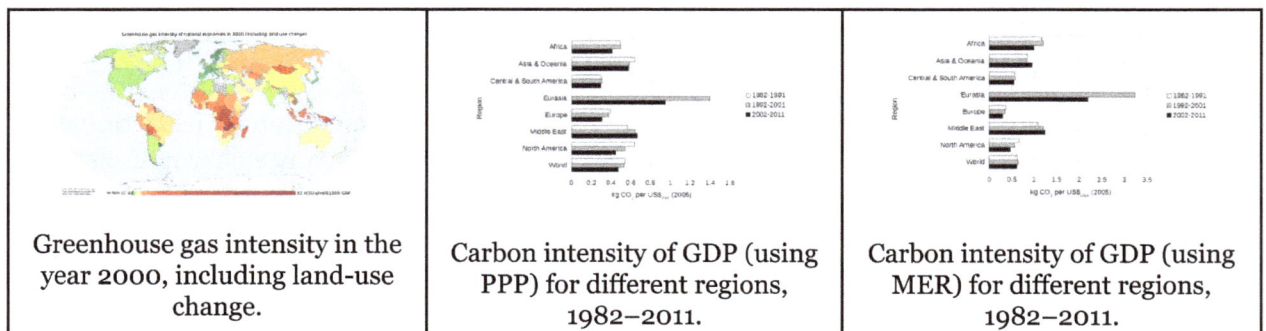

Greenhouse gas intensity in the year 2000, including land-use change.	Carbon intensity of GDP (using PPP) for different regions, 1982–2011.	Carbon intensity of GDP (using MER) for different regions, 1982–2011.

Greenhouse gas intensity is a ratio between greenhouse gas emissions and another metric, e.g., gross domestic product (GDP) or energy use. The terms "carbon intensity" and "emissions intensity" are also sometimes used. GHG intensities may be calculated using market exchange rates (MER) or purchasing power parity (PPP) (Banuri *et al.*, 1996, p. 96). Calculations based on MER show large differences in intensities between developed and developing countries, whereas calculations based on PPP show smaller differences.

Cumulative and Historical Emissions

Cumulative energy-related CO_2 emissions between the years 1850–2005 grouped into low-income, middle-income, high-income, the EU-15, and the OECD countries.

Cumulative energy-related CO_2 emissions between the years 1850–2005 for individual countries.

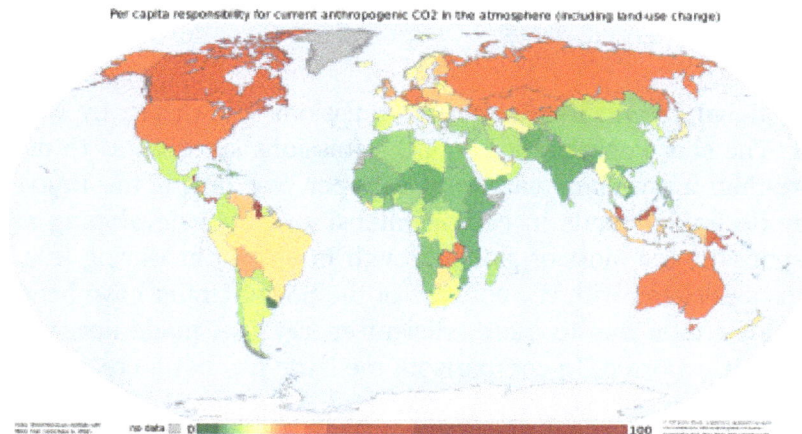

Map of cumulative per capita anthropogenic atmospheric CO_2 emissions by country. Cumulative emissions include land use change, and are measured between the years 1950 and 2000.

Regional trends in annual CO_2 emissions from fuel combustion between 1971 and 2009.

Regional trends in annual per capita CO_2 emissions from fuel combustion between 1971 and 2009.

Cumulative anthropogenic (i.e., human-emitted) emissions of CO_2 from fossil fuel use are a major cause of global warming, and give some indication of which countries have contributed most to human-induced climate change.

Top-5 historic CO_2 contributors by region over the years 1800 to 1988 (in %)		
Region	**Industrial CO_2**	**Total CO_2**
OECD North America	33.2	29.7
OECD Europe	26.1	16.6
Former USSR	14.1	12.5
China	5.5	6.0
Eastern Europe	5.5	4.8

The table above to the left is based on Banuri *et al.* (1996, p. 94). Overall, developed countries accounted for 83.8% of industrial CO_2 emissions over this time period, and 67.8% of total CO_2 emissions. Developing countries accounted for industrial CO_2 emissions of 16.2% over this time period, and 32.2% of total CO_2 emissions. The estimate of total CO_2 emissions includes biotic carbon emissions, mainly from deforestation. Banuri *et al.* (1996, p. 94) calculated per capita cumulative emissions based on then-current population. The ratio in per capita emissions between industrialized countries and developing countries was estimated at more than 10 to 1.

Including biotic emissions brings about the same controversy mentioned earlier regarding carbon sinks and land-use change (Banuri *et al.*, 1996, pp. 93–94). The actual calculation of net emissions is very complex, and is affected by how carbon sinks are allocated between regions and the dynamics of the climate system.

Non-OECD countries accounted for 42% of cumulative energy-related CO_2 emissions between 1890–2007. Over this time period, the US accounted for 28% of emissions; the EU, 23%; Russia, 11%; China, 9%; other OECD countries, 5%; Japan, 4%; India, 3%; and the rest of the world, 18%.

Changes Since a Particular Base Year

Between 1970–2004, global growth in annual CO_2 emissions was driven by North America, Asia, and the Middle East. The sharp acceleration in CO_2 emissions since 2000 to more than a 3% increase per year (more than 2 ppm per year) from 1.1% per year during the 1990s is attributable to the lapse of formerly declining trends in carbon intensity of both developing and developed nations. China was responsible for most of global growth in emissions during this period. Localised plummeting emissions associated with the collapse of the Soviet Union have been followed by slow emissions growth in this region due to more efficient energy use, made necessary by the increasing proportion of it that is exported. In comparison, methane has not increased appreciably, and N 2O by 0.25% y^{-1}.

Using different base years for measuring emissions has an effect on estimates of national contri-butions to global warming. This can be calculated by dividing a country's highest contribution to global warming starting from a particular base year, by that country's minimum contribution to global warming starting from a particular base year. Choosing between different base years of 1750, 1900, 1950, and 1990 has a significant effect for most countries. Within the G8 group of countries, it is most significant for the UK, France and Germany. These countries have a long history of CO_2 emissions.

Annual Emissions

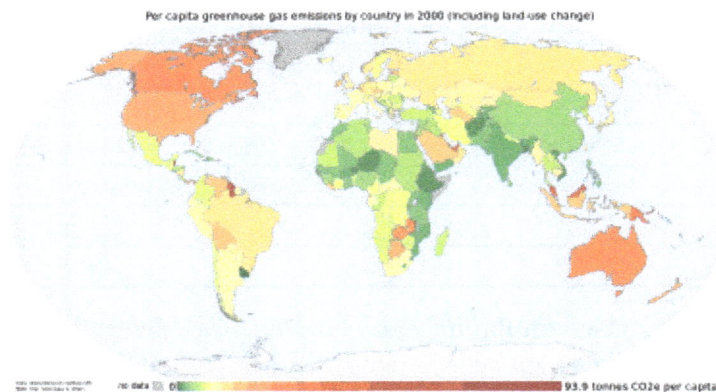

Per capita anthropogenic greenhouse gas emissions by country for the year 2000 including land-use change.

Annual per capita emissions in the industrialized countries are typically as much as ten times the average in developing countries. Due to China's fast economic development, its annual per capita emissions are quickly approaching the levels of those in the Annex I group of the Kyoto Protocol (i.e., the developed countries excluding the USA). Other countries with fast growing emissions are South Korea, Iran, and Australia (which apart from the oil rich Persian Gulf states, now has the highest percapita emission rate in the world). On the other hand, annual per capita emissions of the EU-15 and the USA are gradually decreasing over time. Emissions in Russia and Ukraine have decreased fastest since 1990 due to economic restructuring in these countries.

Energy statistics for fast growing economies are less accurate than those for the industrialized countries. For China's annual emissions in 2008, the Netherlands Environmental Assessment Agency estimated an uncertainty range of about 10%.

The GHG footprint, or greenhouse gas footprint, refers to the amount of GHG that are emitted during the creation of products or services. It is more comprehensive than the commonly used carbon footprint, which measures only carbon dioxide, one of many greenhouse gases.

2015 was the first year to see both total global economic growth and a reduction of carbon emissions.

Top Emitter Countries

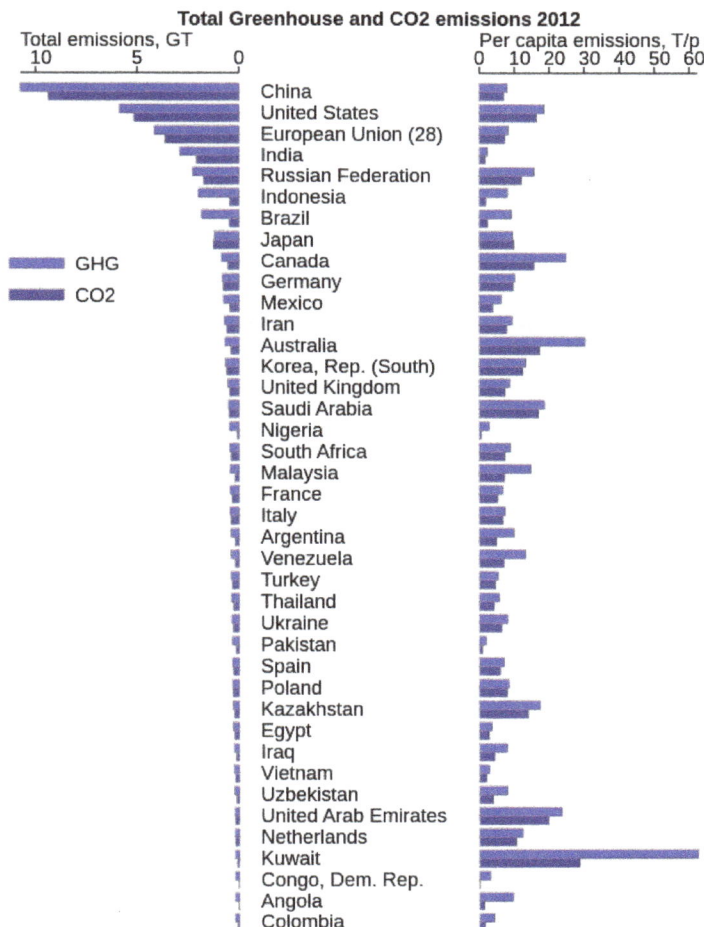

The top 40 countries emitting all greenhouse gases, showing both that derived from all sources including land clearance and forestry and also the CO2 component excluding those sources. Per capita figures are included. Data taken from World Resources Institute, Washington. Note that Indonesia and Brazil show very much higher than on graphs simply showing fossil fuel use.

Annual

In 2009, the annual top ten emitting countries accounted for about two-thirds of the world's annual energy-related CO_2 emissions.

Top-10 annual energy-related CO_2 emitters for the year 2009		
Country	% of global total annual emissions	Tonnes of GHG per capita
People's Rep. of China	23.6	5.1
United States	17.9	16.9
India	5.5	1.4
Russian Federation	5.3	10.8
Japan	3.8	8.6
Germany	2.6	9.2
Islamic Rep. of Iran	1.8	7.3
Canada	1.8	15.4
South Korea	1.8	10.6
United Kingdom	1.6	7.5

Cumulative

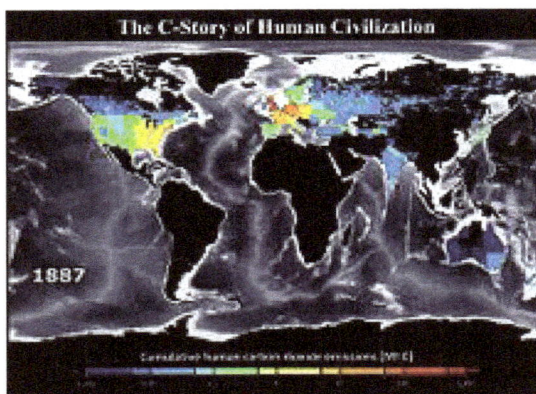

The C-Story of Human Civilization by PIK

Top-10 cumulative energy-related CO_2 emitters between 1850–2008		
Country	% of world total	Metric tonnes CO_2 per person
United States	28.5	1,132.7
China	9.36	85.4
Russian Federation	7.95	677.2
Germany	6.78	998.9
United Kingdom	5.73	1,127.8
Japan	3.88	367
France	2.73	514.9
India	2.52	26.7
Canada	2.17	789.2
Ukraine	2.13	556.4

Embedded Emissions

One way of attributing greenhouse gas (GHG) emissions is to measure the embedded emissions (also referred to as "embodied emissions") of goods that are being consumed. Emissions are usually measured according to production, rather than consumption. For example, in the main international treaty on climate change (the UNFCCC), countries report on emissions produced within their borders, e.g., the emissions produced from burning fossil fuels. Under a production-based account-ing of emissions, embedded emissions on imported goods are attributed to the exporting, rather than the importing, country. Under a consumption-based accounting of emissions, embedded emissions on imported goods are attributed to the importing country, rather than the exporting, country.

Davis and Caldeira (2010) found that a substantial proportion of CO_2 emissions are traded internationally. The net effect of trade was to export emissions from China and other emerging markets to consumers in the US, Japan, and Western Europe. Based on annual emissions data from the year 2004, and on a per-capita consumption basis, the top-5 emitting countries were found to be (in tCO_2 per person, per year): Luxembourg (34.7), the US (22.0), Singapore (20.2), Australia (16.7), and Canada (16.6). Carbon Trust research revealed that approximately 25% of all CO_2 emissions from human activities 'flow' (i.e. are imported or exported) from one country to another. Major developed economies were found to be typically net importers of embodied carbon emissions — with UK consumption emissions 34% higher than production emissions, and Germany (29%), Japan (19%) and the USA (13%) also significant net importers of embodied emissions.

Effect of Policy

Governments have taken action to reduce GHG emissions (climate change mitigation). Assessments of policy effectiveness have included work by the Intergovernmental Panel on Climate Change, International Energy Agency, and United Nations Environment Programme. Policies implemented by governments have included national and regional targets to reduce emissions, promoting energy efficiency, and support for renewable energy such as Solar energy as an effective use of renewable energy because solar uses energy from the sun and does not release pollutants into the air.

Countries and regions listed in Annex I of the United Nations Framework Convention on Climate Change (UNFCCC) (i.e., the OECD and former planned economies of the Soviet Union) are required to submit periodic assessments to the UNFCCC of actions they are taking to address climate change. Analysis by the UNFCCC (2011) suggested that policies and measures undertaken by Annex I Parties may have produced emission savings of 1.5 thousand Tg CO_2-eq in the year 2010, with most savings made in the energy sector. The projected emissions saving of 1.5 thousand Tg CO_2-eq is measured against a hypothetical "baseline" of Annex I emissions, i.e., projected Annex I emissions in the absence of policies and measures. The total projected Annex I saving of 1.5 thou-sand CO_2-eq does not include emissions savings in seven of the Annex I Parties.

Projections

A wide range of projections of future GHG emissions have been produced. Rogner *et al.* (2007) assessed the scientific literature on GHG projections. Rogner *et al.* (2007) concluded that unless energy policies changed substantially, the world would continue to depend on fossil fuels until 2025–

2030. Projections suggest that more than 80% of the world's energy will come from fossil fuels. This conclusion was based on "much evidence" and "high agreement" in the literature. Projected annual energy-related CO_2 emissions in 2030 were 40–110% higher than in 2000, with two-thirds of the increase originating in developing countries. Projected annual per capita emissions in developed country regions remained substantially lower (2.8–5.1 tonnes CO_2) than those in developed country regions (9.6–15.1 tonnes CO_2). Projections consistently showed increase in annual world GHG emissions (the "Kyoto" gases, measured in CO_2-equivalent) of 25–90% by 2030, compared to 2000.

Relative CO_2 Emission from Various Fuels

One liter of gasoline, when used as a fuel, produces 2.32 kg (about 1300 liters or 1.3 cubic meters) of carbon dioxide, a greenhouse gas. One US gallon produces 19.4 lb (1,291.5 gallons or 172.65 cubic feet)

Mass of carbon dioxide emitted per quantity of energy for various fuels			
Fuel name	CO_2 emitted (lbs/10^6 Btu)	CO_2 emitted (g/MJ)	CO_2 emitted (g/KWh)
Natural gas	117	50.30	181.08
Liquefied petroleum gas	139	59.76	215.14
Propane	139	59.76	215.14
Aviation gasoline	153	65.78	236.81
Automobile gasoline	156	67.07	241.45
Kerosene	159	68.36	246.10
Fuel oil	161	69.22	249.19
Tires/tire derived fuel	189	81.26	292.54
Wood and wood waste	195	83.83	301.79
Coal (bituminous)	205	88.13	317.27
Coal (sub-bituminous)	213	91.57	329.65
Coal (lignite)	215	92.43	332.75
Petroleum coke	225	96.73	348.23
Tar-sand Bitumen			
Coal (anthracite)	227	97.59	351.32

Life-Cycle Greenhouse-Gas Emissions of Energy Sources

A literature review of numerous energy sources CO_2 emissions by the IPCC in 2011, found that, the CO_2 emission value that fell within the 50th percentile of all total life cycle emissions studies conducted was as follows.

Lifecycle greenhouse gas emissions by electricity source.		
Technology	Description	50th percentile (g CO_2/kWh$_e$)
Hydroelectric	reservoir	4
Ocean Energy	wave and tidal	8
Wind	onshore	12

Nuclear	various generation II reactor types	16
Biomass	various	18
Solar thermal	parabolic trough	22
Geothermal	hot dry rock	45
Solar PV	Polycrystalline silicon	46
Natural gas	various combined cycle turbines without scrubbing	469
Coal	various generator types without scrubbing	1001

Removal from the Atmosphere ("Sinks")

Natural Processes

Greenhouse gases can be removed from the atmosphere by various processes, as a consequence of:

- a physical change (condensation and precipitation remove water vapor from the atmosphere).

- a chemical reaction within the atmosphere. For example, methane is oxidized by reaction with naturally occurring hydroxyl radical, OH· and degraded to CO_2 and water vapor (CO_2 from the oxidation of methane is not included in the methane Global warming potential). Other chemical reactions include solution and solid phase chemistry occurring in atmospheric aerosols.

- a physical exchange between the atmosphere and the other compartments of the planet. An example is the mixing of atmospheric gases into the oceans.

- a chemical change at the interface between the atmosphere and the other compartments of the planet. This is the case for CO_2, which is reduced by photosynthesis of plants, and which, after dissolving in the oceans, reacts to form carbonic acid and bicarbonate and carbonate ions.

- a photochemical change. Halocarbons are dissociated by UV light releasing Cl· and F· as free radicals in the stratosphere with harmful effects on ozone (halocarbons are generally too stable to disappear by chemical reaction in the atmosphere).

Negative Emissions

A number of technologies remove greenhouse gases emissions from the atmosphere. Most widely analysed are those that remove carbon dioxide from the atmosphere, either to geologic formations such as bio-energy with carbon capture and storage and carbon dioxide air capture, or to the soil as in the case with biochar. The IPCC has pointed out that many long-term climate scenario models require large scale manmade negative emissions to avoid serious climate change.

History of Scientific Research

In the late 19th century scientists experimentally discovered that N2 and O2 do not absorb infrared radiation (called, at that time, "dark radiation"), while water (both as true vapor and condensed in the form of microscopic droplets suspended in clouds) and CO_2 and other poly-atomic gaseous molecules do absorb infrared radiation. In the early 20th century researchers realized that green-house gases in the atmosphere made Earth's overall temperature higher than it would be without them. During the late 20th century, a scientific consensus evolved that increasing concentrations of greenhouse gases in the atmosphere cause a substantial rise in global temperatures and changes to other parts of the climate system, with consequences for the environment and for human health.

Water Vapor

Water vapor, water vapour or aqueous vapor, is the gaseous phase of water. It is one state of water within the hydrosphere. Water vapor can be produced from the evaporation or boiling of liquid water or from the sublimation of ice. Unlike other forms of water, water vapor is invisible. Under typical atmospheric conditions, water vapor is continuously generated by evaporation and removed by condensation. It is lighter than air and triggers convection currents that can lead to clouds.

Being a component of Earth's hydrosphere and hydrologic cycle, it is particularly abundant in Earth's atmosphere where it is also a potent greenhouse gas along with other gases such as carbon dioxide and methane. Use of water vapor, as steam, has been important to humans for cooking and as a major component in energy production and transport systems since the industrial revolution.

Water vapor is a relatively common atmospheric constituent, present even in the solar atmosphere as well as every planet in the Solar System and many astronomical objects including natural satellites, comets and even large asteroids. Likewise the detection of extrasolar water vapor would indicate a similar distribution in other planetary systems. Water vapor is significant in that it can be indirect evidence supporting the presence of extraterrestrial liquid water in the case of some planetary mass objects.

Properties

Evaporation

Whenever a water molecule leaves a surface and diffuses into a surrounding gas, it is said to have evaporated. Each individual water molecule which transitions between a more associated (liquid) and a less associated (vapor/gas) state does so through the absorption or release of kinetic energy. The aggregate measurement of this kinetic energy transfer is defined as thermal energy and occurs only when there is differential in the temperature of the water molecules. Liquid water that becomes water vapor takes a parcel of heat with it, in a process called evaporative cooling. The amount of water vapor in the air determines how fast each molecule will return to the surface. When a net evaporation occurs, the body of water will undergo a net cooling directly related to the loss of water.

In the US, the National Weather Service measures the actual rate of evaporation from a standardized "pan" open water surface outdoors, at various locations nationwide. Others do likewise around the world. The US data is collected and compiled into an annual evaporation map. The measurements range from under 30 to over 120 inches per year. Formulas can be used for calculating the rate of evaporation from a water surface such as a swimming pool. In some countries, the evaporation rate far exceeds the precipitation rate.

Evaporative cooling is restricted by atmospheric conditions. Humidity is the amount of water vapor in the air. The vapor content of air is measured with devices known as hygrometers. The measurements are usually expressed as specific humidity or percent relative humidity. The temperatures of the atmosphere and the water surface determine the equilibrium vapor pressure; 100% relative humidity occurs when the partial pressure of water vapor is equal to the equilibrium vapor pressure. This condition is often referred to as complete saturation. Humidity ranges from 0 gram per cubic metre in dry air to 30 grams per cubic metre (0.03 ounce per cubic foot) when the vapor is saturated at 30 °C.

Recovery of meteorites in Antarctica (ANSMET)

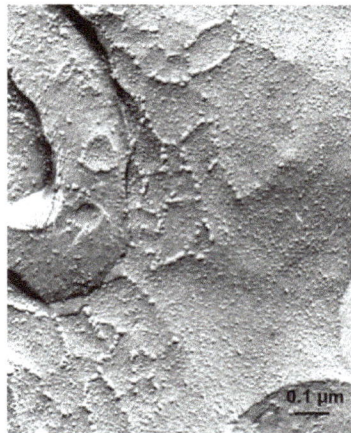

Electron micrograph of freeze-etched capillary tissue

Sublimation

Another form of evaporation is sublimation, by which water molecules become gaseous directly, leaving the surface of ice without first becoming liquid water. Sublimation accounts for the slow mid-winter disappearance of ice and snow at temperatures too low to cause melting. Antarctica shows this effect to a unique degree because it is by far the continent with the lowest rate of precipitation on Earth. As a result, there are large areas where millennial layers of snow have sublimed, leaving behind whatever non-volatile materials they had contained. This is extremely valuable to certain scientific disciplines, a dramatic example being the collection of meteorites that are left exposed in unparalleled numbers and excellent states of preservation.

Sublimation is of importance in the preparation of certain classes of biological specimens for scanning electron microscopy. Typically the specimens are prepared by cryofixation and freeze-fracture, after which the broken surface is freeze-etched, being eroded by exposure to vacuum till it shows the required level of detail. This technique can display protein molecules, organelle structures and lipid bilayers with very low degrees of distortion.

Condensation

Clouds, formed by condensed water vapor

Water vapor will only condense onto another surface when that surface is cooler than the dew point temperature, or when the water vapor equilibrium in air has been exceeded. When water vapor condenses onto a surface, a net warming occurs on that surface. The water molecule brings heat energy with it. In turn, the temperature of the atmosphere drops slightly. In the atmosphere, condensation produces clouds, fog and precipitation (usually only when facilitated by cloud condensation nuclei). The dew point of an air parcel is the temperature to which it must cool before water vapor in the air begins to condense.

Also, a net condensation of water vapor occurs on surfaces when the temperature of the surface is at or below the dew point temperature of the atmosphere. Deposition is a phase transition separate from condensation which leads to the direct formation of ice from water vapor. Frost and snow are examples of deposition.

Chemical Reactions

A number of chemical reactions have water as a product. If the reactions take place at temperatures higher than the dew point of the surrounding air the water will be formed as vapor and increase the local humidity, if below the dew point local condensation will occur. Typical reactions that result in water formation are the burning of hydrogen or hydrocarbons in air or other oxygen containing gas mixtures, or as a result of reactions with oxidizers.

In a similar fashion other chemical or physical reactions can take place in the presence of water vapor resulting in new chemicals forming such as rust on iron or steel, polymerization occurring (certain polyurethane foams and cyanoacrylate glues cure with exposure to atmospheric humidity) or forms changing such as where anhydrous chemicals may absorb enough vapor to form a crystalline structure or alter an existing one, sometimes resulting in characteristic color changes that can be used for measurement.

Measurement

Measuring the quantity of water vapor in a medium can be done directly or remotely with varying degrees of accuracy. Remote methods such electromagnetic absorption are possible from satellites above planetary atmospheres. Direct methods may use electronic transducers, moistened thermometers or hygroscopic materials measuring changes in physical properties or dimensions.

	medium	temperature range (degC)	measurement uncertainty	typical measurement frequency	system cost	notes
sling psychrometer	air	−10 to 50	low to moderate	hourly	low	
satellite-based spectroscopy	air	−80 to 60	low		very high	
capacitive sensor	air/gases	−40 to 50	moderate	2 to 0.05 Hz	medium	prone to becoming saturated/contaminated over time
warmed capacitive sensor	air/gases	−15 to 50	moderate to low	2 to 0.05 Hz (temp dependant)	medium to high	prone to becoming saturated/contaminated over time
resistive sensor	air/gases	−10 to 50	moderate	60 seconds	medium	prone to contamination
lithium chloride dewcell	air	−30 to 50	moderate	continuous	medium	see dewcell
Cobalt(II) chloride	air/gases	0 to 50	high	5 minutes	very low	often used in Humidity indicator card
Absorption spectroscopy	air/gases		moderate		high	
Aluminum oxide	air/gases		moderate		medium	see Moisture analysis
silicon oxide	air/gases		moderate		medium	see Moisture analysis
Piezoelectric sorption	air/gases		moderate		medium	see Moisture analysis
Electrolytic	air/gases		moderate		medium	see Moisture analysis

hair tension	air	0 to 40	high	continuous	low to medium	Affected by temperature. Adversely affected by prolonged high concentrations
Nephelome-ter	air/oth-er gases		low		very high	
Goldbeater's skin (cow Peritoneum)	air	−20 to 30	moderate (with cor-rections)	slow, slower at lower tempera-tures	low	ref:WMO Guide to Me-teorological Instruments and Methods of Observa-tion No. 8 2006, (pages 1.12−1)
Lyman-alpha				high fre-quency	high	http://amsglossary. allenpress.com/glossary/ search?id=lyman-al-pha-hygrometer1 Re-quires frequent calibra-tion
Gravimetric Hygrometer			very low		very high	often called primary source, national indepen-dent standards developed in US,UK,EU & Japan

Impact on Air Density

Water vapor is lighter or less dense than dry air. At equivalent temperatures it is buoyant with respect to dry air, whereby the density of dry air at standard temperature and pressure is 1.27 g/L and water vapor at standard temperature and pressure has the much lower density of 0.804 g/L.

Calculations

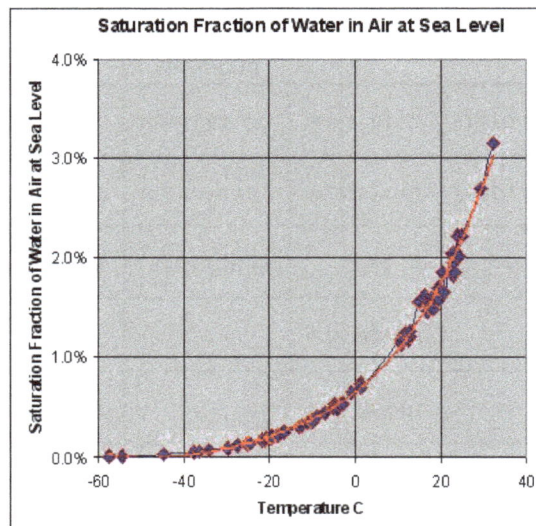

Water vapor and dry air density calculations at 0 °C:

- The molar mass of water is 18.02 g/mol, as calculated from the sum of the atomic masses of its constituent atoms.

- The average molecular mass of air (approx. 78% nitrogen, N_2; 21% oxygen, O_2; 1% other gases) is 28.57 g/mol at standard temperature and pressure (STP).

- Using Avogadro's Law and the ideal gas law, water vapor and air will have a molar volume of 22.414 L/mol at STP. A molar mass of air and water vapor occupy the same volume of 22.414 litres. The density (mass/volume) of water vapor is 0.804 g/L, which is significantly less than that of dry air at 1.27 g/L at STP. This means water vapor is lighter than air.

- STP conditions imply a temperature of 0 °C, at which the ability of water to become vapor is very restricted. Its concentration in air is very low at 0 °C. The red line on the chart to the right is the maximum concentration of water vapor expected for a given temperature. The water vapor concentration increases significantly as the temperature rises, approaching 100% (steam, pure water vapor) at 100 °C. However the difference in densities between air and water vapor would still exist.

At Equal Temperatures

At the same temperature, a column of dry air will be denser or heavier than a column of air containing any water vapor, the molar mass of diatomic nitrogen and diatomic oxygen both being greater than the molar mass of water. Thus, any volume of dry air will sink if placed in a larger volume of moist air. Also, a volume of moist air will rise or be buoyant if placed in a larger region of dry air. As the temperature rises the proportion of water vapor in the air increases, and its buoyancy will increase. The increase in buoyancy can have a significant atmospheric impact, giving rise to powerful, moisture rich, upward air currents when the air temperature and sea temperature reaches 25 °C or above. This phenomenon provides a significant driving force for cyclonic and anticyclonic weather systems (typhoons and hurricanes).

Respiration and Breathing

Water vapor is a by-product of respiration in plants and animals. Its contribution to the pressure, increases as its concentration increases. Its partial pressure contribution to air pressure increases, lowering the partial pressure contribution of the other atmospheric gases (Dalton's Law). The total air pressure must remain constant. The presence of water vapor in the air naturally dilutes or displaces the other air components as its concentration increases.

This can have an effect on respiration. In very warm air (35 °C) the proportion of water vapor is large enough to give rise to the stuffiness that can be experienced in humid jungle conditions or in poorly ventilated buildings.

Lifting Gas

Water vapor has lower density than that of air and is therefore buoyant in air but has lower vapor pressure than that of air. When water vapor is used as a lifting gas for use by a thermal airship the water vapor is heated to form steam so that its vapor pressure is greater than the surrounding air pressure in order to pressurize and to maintain the shape a theoretical "steam balloon", which yields approximately 60% the lift of helium and twice that of hot air.

General Discussion

The amount of water vapor in an atmosphere is constrained by the restrictions of partial pressures and temperature. Dew point temperature and relative humidity act as guidelines for the process of water vapor in the water cycle. Energy input, such as sunlight, can trigger more evaporation on an ocean surface or more sublimation on a chunk of ice on top of a mountain. The *balance* between condensation and evaporation gives the quantity called vapor partial pressure.

The maximum partial pressure (*saturation pressure*) of water vapor in air varies with temperature of the air and water vapor mixture. A variety of empirical formulas exist for this quantity; the most used reference formula is the Goff-Gratch equation for the SVP over liquid water below zero degree Celsius:

$$\log_{10}(p) = -7.90298\left(\frac{373.16}{T} - 1\right) + 5.02808 \log_{10} \frac{373.16}{T}$$

$$+8.1328 \times 10^{-3}\left(10^{-3.49149\left(\frac{373.16}{T}-1\right)} - 1\right)$$

$$+8.1328 \times 10^{-3}\left(10^{-3.49149\left(\frac{373.16}{T}-1\right)} - 1\right)$$

$$+\log_{10}(1013.246)$$

Where T, temperature of the moist air, is given in units of kelvin, and p is given in units of millibars (hectopascals).

The formula is valid from about −50 to 102 °C; however there are a very limited number of measurements of the vapor pressure of water over supercooled liquid water. There are a number of other formulae which can be used.

Under certain conditions, such as when the boiling temperature of water is reached, a net evaporation will always occur during standard atmospheric conditions regardless of the percent of relative humidity. This immediate process will dispel massive amounts of water vapor into a cooler atmosphere.

Exhaled air is almost fully at equilibrium with water vapor at the body temperature. In the cold air the exhaled vapor quickly condenses, thus showing up as a fog or mist of water droplets and as condensation or frost on surfaces. Forcibly condensing these water droplets from exhaled breath is the basis of exhaled breath condensate, an evolving medical diagnostic test.

Controlling water vapor in air is a key concern in the heating, ventilating, and air-conditioning (HVAC) industry. Thermal comfort depends on the moist air conditions. Non-human comfort situations are called refrigeration, and also are affected by water vapor. For example, many food stores, like supermarkets, utilize open chiller cabinets, or *food cases*, which can significantly lower the water vapor pressure (lowering humidity). This practice delivers several benefits as well as problems.

In Earth's Atmosphere

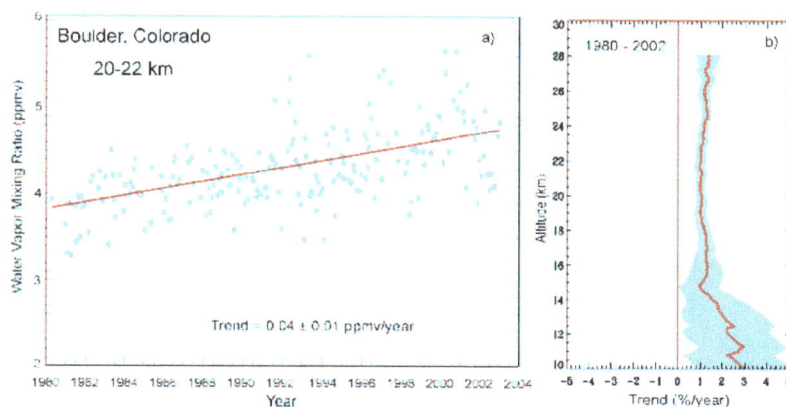

Evidence for increasing amounts of stratospheric water vapor over time in Boulder, Colorado.

Gaseous water represents a small but environmentally significant constituent of the atmosphere. The percentage water vapor in surface air varies from 0.01% at -42 °C (-44 °F) to 4.24% when the dew point is 30 °C (86 °F). Approximately 99.13% of it is contained in the troposphere. The condensation of water vapor to the liquid or ice phase is responsible for clouds, rain, snow, and other precipitation, all of which count among the most significant elements of what we experience as weather. Less obviously, the latent heat of vaporization, which is released to the atmosphere whenever condensation occurs, is one of the most important terms in the atmospheric energy budget on both local and global scales. For example, latent heat release in atmospheric convection is directly responsible for powering destructive storms such as tropical cyclones and severe thunderstorms. Water vapor is also the most potent greenhouse gas owing to the presence of the hydroxyl bond which strongly absorbs in the infra-red region of the light spectrum.

Water in Earth's atmosphere is not merely below its boiling point (100 °C), but at altitude it goes below its freezing point (0 °C), due to water's highly polar attraction. When combined with its quantity, water vapor then has a relevant dew point and frost point, unlike e. g., carbon dioxide and methane. Water vapor thus has a scale height a fraction of that of the bulk atmosphere, as the water condenses and exits, primarily in the troposphere, the lowest layer of the atmosphere. Carbon dioxide (CO_2) and methane, being non-polar, rise above water vapor. The absorption and emission of both compounds contribute to Earth's emission to space, and thus the planetary greenhouse effect. This greenhouse forcing is directly observable, via distinct spectral features versus water vapor, and observed to be rising with rising CO_2 levels. Conversely, adding water vapor at high altitudes has a disproportionate impact, which is why methane (rising, then oxidizing to CO_2 and two water molecules) and jet traffic have disproportionately high warming effects.

It is less clear how cloudiness would respond to a warming climate; depending on the nature of the response, clouds could either further amplify or partly mitigate warming from long-lived greenhouse gases.

In the absence of other greenhouse gases, Earth's water vapor would condense to the surface; this has likely happened, possibly more than once. Scientists thus distinguish between non-condensable (driving) and condensable (driven) greenhouse gases- i. e., the above water vapor feedback.

Fog and clouds form through condensation around cloud condensation nuclei. In the absence of nuclei, condensation will only occur at much lower temperatures. Under persistent condensation or deposition, cloud droplets or snowflakes form, which precipitate when they reach a critical mass.

The water content of the atmosphere as a whole is constantly depleted by precipitation. At the same time it is constantly replenished by evaporation, most prominently from seas, lakes, rivers, and moist earth. Other sources of atmospheric water include combustion, respiration, volcanic eruptions, the transpiration of plants, and various other biological and geological processes. The mean global content of water vapor in the atmosphere is roughly sufficient to cover the surface of the planet with a layer of liquid water about 25 mm deep. The mean annual precipitation for the planet is about 1 meter, which implies a rapid turnover of water in the air – on average, the residence time of a water molecule in the troposphere is about 9 to 10 days.

Episodes of surface geothermal activity, such as volcanic eruptions and geysers, release variable amounts of water vapor into the atmosphere. Such eruptions may be large in human terms, and major explosive eruptions may inject exceptionally large masses of water exceptionally high into the atmosphere, but as a percentage of total atmospheric water, the role of such processes is minor. The relative concentrations of the various gases emitted by volcanoes varies considerably according to the site and according to the particular event at any one site. However, water vapor is consistently the commonest volcanic gas; as a rule, it comprises more than 60% of total emissions during a subaerial eruption.

Atmospheric water vapor content is expressed using various measures. These include vapor pressure, specific humidity, mixing ratio, dew point temperature, and relative humidity.

Radar and Satellite Imaging

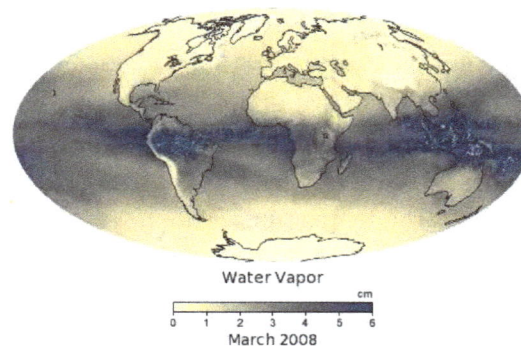

These maps show the average amount of water vapor in a column of atmosphere in a given month.

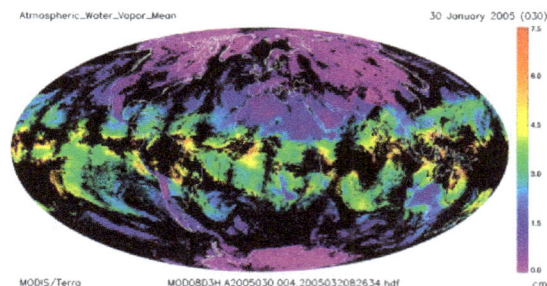

MODIS/Terra global mean atmospheric water vapor

Because water molecules absorb microwaves and other radio wave frequencies, water in the atmosphere attenuates radar signals. In addition, atmospheric water will reflect and refract signals to an extent that depends on whether it is vapor, liquid or solid.

Generally, radar signals lose strength progressively the farther they travel through the troposphere. Different frequencies attenuate at different rates, such that some components of air are opaque to some frequencies and transparent to others. Radio waves used for broadcasting and other communication experience the same effect.

Water vapor reflects radar to a lesser extent than do water's other two phases. In the form of drops and ice crystals, water acts as a prism, which it does not do as an individual molecule; however, the existence of water vapor in the atmosphere causes the atmosphere to act as a giant prism.

A comparison of GOES-12 satellite images shows the distribution of atmospheric water vapor relative to the oceans, clouds and continents of the Earth. Vapor surrounds the planet but is unevenly distributed. The image loop on the right shows monthly average of water vapor content with the units are given in centimeters, which is the precipitable water or equivalent amount of water that could be produced if all the water vapor in the column were to condense. The lowest amounts of water vapor (0 centimeters) appear in yellow, and the highest amounts (6 centimeters) appear in dark blue. Areas of missing data appear in shades of gray. The maps are based on data collected by the Moderate Resolution Imaging Spectroradiometer (MODIS) sensor on NASA's Aqua satellite. The most noticeable pattern in the time series is the influence of seasonal temperature changes and incoming sunlight on water vapor. In the tropics, a band of extremely humid air wobbles north and south of the equator as the seasons change. This band of humidity is part of the Intertropical Convergence Zone, where the easterly trade winds from each hemisphere converge and produce near-daily thunderstorms and clouds. Farther from the equator, water vapor concentrations are high in the hemisphere experiencing summer and low in the one experiencing winter. Another pattern that shows up in the time series is that water vapor amounts over land areas decrease more in winter months than adjacent ocean areas do. This is largely because air temperatures over land drop more in the winter than temperatures over the ocean. Water vapor condenses more rapidly in colder air.

As water vapour absorbs light in the visible spectral range, its absorption can be used in spectroscopic applications (such as DOAS) to determine the amount of water vapor in the atmosphere. This is done operationally, e.g. from the GOME spectrometers on ERS and MetOp. The weaker water vapor absorption lines in the blue spectral range and further into the UV up to its dissociation limit around 243 nm are mostly based on quantum mechanical calculations and are only partly confirmed by experiments.

Lightning Generation

Water vapor plays a key role in lightning production in the atmosphere. From cloud physics, usually, clouds are the real generators of static charge as found in Earth's atmosphere. But the ability, or capability of clouds to hold massive amounts of electrical energy is directly related to the amount of water vapor present in the local system.

The amount of water vapor directly controls the permittivity of the air. During times of low humidity, static discharge is quick and easy. During times of higher humidity, fewer static discharges occur. Permittivity and capacitance work hand in hand to produce the megawatt outputs of lightning.

After a cloud, for instance, has started its way to becoming a lightning generator, atmospheric water vapor acts as a substance (or insulator) that decreases the ability of the cloud to discharge its electrical energy. Over a certain amount of time, if the cloud continues to generate and store more static electricity, the barrier that was created by the atmospheric water vapor will ultimately break down from the stored electrical potential energy. This energy will be released to a locally, oppositely charged region in the form of lightning. The strength of each discharge is directly related to the atmospheric permittivity, capacitance, and the source's charge generating ability.

In the Solar System

Water vapor is a significant component of the Earth's atmosphere and a greenhouse gas. It is also common in the Solar System and by extension, other planetary systems. Its signature has been detected in the atmospheres of the Sun, occurring in sunspots. The presence of water vapor has been detected in the atmospheres of Mercury, Venus, Earth (and Moon), Mars, Jupiter, Saturn, Uranus and Neptune, the planets of the Solar System, although typically in only trace amounts.

Cryogeyser erupting on Jupiter's moon Europa (artist concept)

Artist's illustration of the signatures of water in exoplanet atmospheres detectable by instruments such as the Hubble Space Telescope.

Geological formations such as cryogeysers are thought to exist on the surface of several icy moons ejecting water vapor due to tidal heating and may indicate the presence of substantial quantities of subsurface water. Plumes of water vapor have been detected on Jupiter's moon Europa and are similar to plumes of water vapor detected on Saturn's moon Enceladus. Traces of water vapor have also been detected in the stratosphere of Titan. Water vapor has been found to be a major constituent of the atmosphere of dwarf planet, Ceres, largest object in the asteroid belt The detection was made by using the far-infrared abilities of the Herschel Space Observatory. The finding is unexpected because comets, not asteroids, are typically considered to "sprout jets and plumes." According to one of the scientists, "The lines are becoming more and more blurred between comets and asteroids." Scientists studying Mars hypothesize that if water moves about the planet, it does so as vapor.

The brilliance of comet tails comes largely from water vapor. On approach to the Sun, the ice many comets carry sublimates to vapor, which reflects light from the Sun. Knowing a comet's distance from the sun, astronomers may deduce a comet's water content from its brilliance.

Water vapor has also been confirmed outside the Solar System. Spectroscopic analysis of HD 209458 b, an extrasolar planet in the constellation Pegasus, provides the first evidence of atmospheric water vapor beyond the Solar System. A star called CW Leonis was found to have a ring of vast quantities of water vapor circling the aging, massive star. A NASA satellite designed to study chemicals in interstellar gas clouds, made the discovery with an onboard spectrometer. Most likely, "the water vapor was vaporized from the surfaces of orbiting comets." HAT-P-11b a relatively small exoplanet has also been found to possess water vapour.

Carbon Dioxide

Carbon dioxide (chemical formula CO_2) is a colorless and odorless gas vital to life on Earth. This naturally occurring chemical compound is composed of a carbon atom covalently double bonded to two oxygen atoms. Carbon dioxide exists in Earth's atmosphere as a trace gas at a concentration of about 0.04 percent (400 ppm) by volume. Natural sources include volcanoes, hot springs and geysers, and it is freed from carbonate rocks by dissolution in water and acids. Because carbon dioxide is soluble in water, it occurs naturally in groundwater, rivers and lakes, in ice caps and glaciers and also in seawater. It is present in deposits of petroleum and natural gas.

Atmospheric carbon dioxide is the primary source of carbon in life on Earth and its concentration in Earth's pre-industrial atmosphere since late in the Precambrian was regulated by photosynthetic organisms and geological phenomena. As part of the carbon cycle, plants, algae, and cyanobacteria use light energy to photosynthesize carbohydrate from carbon dioxide and water, with oxygen produced as a waste product.

Carbon dioxide (CO_2) is produced by all aerobic organisms when they metabolize carbohydrate and lipids to produce energy by respiration. It is returned to water via the gills of fish and to the air via the lungs of air-breathing land animals, including humans. Carbon dioxide is produced during the processes of decay of organic materials and the fermentation of sugars in bread, beer and wine-making. It is produced by combustion of wood, carbohydrates and fossil fuels such as coal, peat, petroleum and natural gas.

It is a versatile industrial material, used, for example, as an inert gas in welding and fire extinguishers, as a pressurizing gas in air guns and oil recovery, as a chemical feedstock and in liquid form as a solvent in decaffeination of coffee and supercritical drying. It is added to drinking water and carbonated beverages including beer and sparkling wine to add effervescence. The frozen solid form of CO_2, known as "dry ice" is used as a refrigerant and as an abrasive in dry-ice blasting.

Carbon dioxide is a significant greenhouse gas. Since the Industrial Revolution, anthropogenic emissions - including the burning of carbon-based fossil fuels and land use changes (primarily deforestation) - have rapidly increased its concentration in the atmosphere, leading to global warming. It is also a major cause of ocean acidification because it dissolves in water to form carbonic acid.

Background

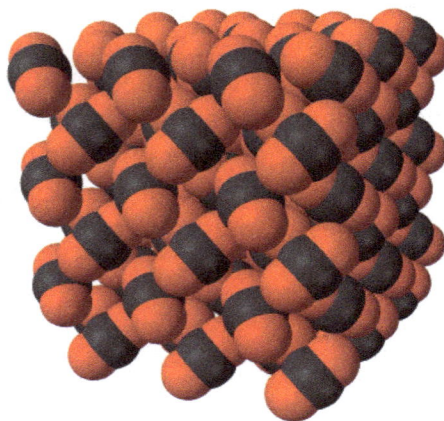

Crystal structure of dry ice

Carbon dioxide was the first gas to be described as a discrete substance. In about 1640, the Flemish chemist Jan Baptist van Helmont observed that when he burned charcoal in a closed vessel, the mass of the resulting ash was much less than that of the original charcoal. His interpretation was that the rest of the charcoal had been transmuted into an invisible substance he termed a "gas" or "wild spirit" (*spiritus sylvestre*).

The properties of carbon dioxide were studied more thoroughly in the 1750s by the Scottish physician Joseph Black. He found that limestone (calcium carbonate) could be heated or treated with acids to yield a gas he called "fixed air." He observed that the fixed air was denser than air and supported neither flame nor animal life. Black also found that when bubbled through limewater (a saturated aqueous solution of calcium hydroxide), it would precipitate calcium carbonate. He used this phenomenon to illustrate that carbon dioxide is produced by animal respiration and microbial fermentation. In 1772, English chemist Joseph Priestley published a paper entitled *Impregnating Water with Fixed Air* in which he described a process of dripping sulfuric acid (or *oil of vitriol* as Priestley knew it) on chalk in order to produce carbon dioxide, and forcing the gas to dissolve by agitating a bowl of water in contact with the gas.

Carbon dioxide was first liquefied (at elevated pressures) in 1823 by Humphry Davy and Michael Faraday. The earliest description of solid carbon dioxide was given by Adrien-Jean-Pierre Thilorier, who in 1835 opened a pressurized container of liquid carbon dioxide, only to find that the cooling produced by the rapid evaporation of the liquid yielded a "snow" of solid CO_2.

Chemical and Physical Properties

$$\ddot{O}\!\!=\!\!C\!\!=\!\!\ddot{O}$$

Carbon Dioxide MO Diagram

Formal Charge	0	0	0
Oxidation State	2–	4+	2–

Stretching and bending oscillations of the CO_2 carbon dioxide molecule. Upper left: symmetric stretching. Upper right: antisymmetric stretching. Lower line: degenerate pair of bending modes.

The carbon dioxide molecule is linear and centrosymmetric. The two C=O bonds are equivalent and are short (116.3 pm), consistent with double bonding. Since it is centrosymmetric, the molecule has no electrical dipole. Consequently, only two vibrational bands are observed in the IR spectrum – an antisymmetric stretching mode at 2349 cm^{-1} and a degenerate pair of bending modes at 667 cm^{-1}. There is also a symmetric stretching mode at 1388 cm^{-1} which is only observed in the Raman spectrum.

In Aqueous Solution

Carbon dioxide is soluble in water, in which it reversibly forms H2CO3 (carbonic acid), which is a weak acid since its ionization in water is incomplete.

CO2 + H2O ⇌ H2CO3

The hydration equilibrium constant of carbonic acid is $K_\mathrm{h} = \dfrac{[H_2CO_3]}{[CO_2(aq)]} = 1.70 \times 10^{-3}\,(at\ 25°C)$.

(at 25 °C). Hence, the majority of the carbon dioxide is not converted into carbonic acid, but remains as CO_2 molecules, not affecting the pH.

The relative concentrations of CO2, H2CO3, and the deprotonated forms HCO−3 (bicarbonate) and CO2−3(carbonate) depend on the pH. As shown in a Bjerrum plot, in neutral or slightly alkaline water (pH > 6.5), the bicarbonate form predominates (>50%) becoming the most prevalent (>95%) at the pH of seawater. In very alkaline water (pH > 10.4), the predominant (>50%) form is carbonate. The oceans, being mildly alkaline with typical pH = 8.2–8.5, contain about 120 mg of bicarbonate per liter.

Being diprotic, carbonic acid has two acid dissociation constants, the first one for the dissociation into the bicarbonate (also called hydrogen carbonate) ion (HCO_3^-):

$$H_2CO_3 \rightleftharpoons HCO_3^- + H^+$$

$K_{a1} = 2.5 \times 10^{-4}$ mol/L; $pK_{a1} = 3.6$ at 25 °C.

This is the *true* first acid dissociation constant, defined as $K_{a1} = \dfrac{[HCO_3^-][H^+]}{[H_2CO_3]}$, where the denominator includes only covalently bound H_2CO_3 and does not include hydrated CO_2(aq). The much smaller and often-quoted value near 4.16×10^{-7} is an *apparent* value calculated on the (incorrect) assumption that all dissolved CO_2 is present as carbonic acid, so that K_{a1}(apparent) $= \dfrac{[HCO_3^-][H^+]}{[H_2CO_3]+[CO_2(aq)]}$ .. Since most of the dissolved CO_2 remains as CO_2 molecules, K_{a1}(apparent) has a much larger denominator and a much smaller value than the true K_{a1}.

The bicarbonate ion is an amphoteric species that can act as an acid or as a base, depending on pH of the solution. At high pH, it dissociates significantly into the carbonate ion (CO_3^{2-}):

$$HCO_3^- \rightleftharpoons CO_3^{2-} + H^+$$

$K_{a2} = 4.69 \times 10^{-11}$ mol/L; $pK_{a2} = 10.329$

In organisms carbonic acid production is catalysed by the enzyme, carbonic anhydrase.

Chemical Reactions of CO_2

CO_2 is a weak electrophile. Its reaction with basic water illustrates this property, in which case hydroxide is the nucleophile. Other nucleophiles react as well. For example, carbanions as provided by Grignard reagents and organolithium compounds react with CO_2 to give carboxylates:

$$MR + CO_2 \rightarrow RCO_2M$$

where M = Li or MgBr and R = alkyl or aryl.

In metal carbon dioxide complexes, CO_2 serves as a ligand, which can facilitate the conversion of CO_2 to other chemicals.

The reduction of CO_2 to CO is ordinarily a difficult and slow reaction:

$$CO_2 + 2\,e^- + 2H^+ \rightarrow CO + H_2O$$

The redox potential for this reaction near pH 7 is about −0.53 V *versus* the standard hydrogen electrode. The nickel-containing enzyme carbon monoxide dehydrogenase catalyses this process.

Physical Properties

Carbon dioxide is colorless. At low concentrations, the gas is odorless. At higher concentrations it has a sharp, acidic odor. At standard temperature and pressure, the density of carbon dioxide is around 1.98 kg/m³, about 1.67 times that of air.

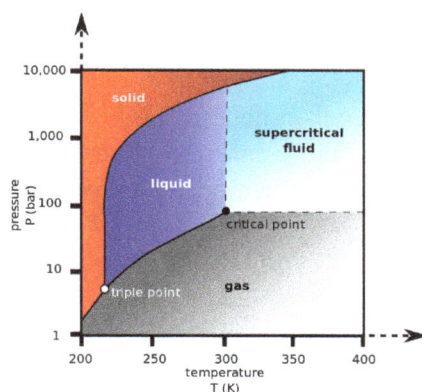

Carbon dioxide pressure-temperature phase diagram showing the triple point and critical point of carbon dioxide

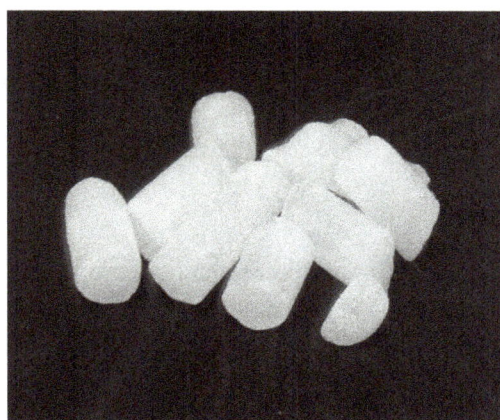

Sample of solid carbon dioxide or "dry ice" pellets

Carbon dioxide has no liquid state at pressures below 5.1 standard atmospheres (520 kPa). At 1 atmosphere (near mean sea level pressure), the gas deposits directly to a solid at temperatures below −78.5 °C (−109.3 °F; 194.7 K) and the solid sublimes directly to a gas above −78.5 °C. In its solid state, carbon dioxide is commonly called dry ice.

Liquid carbon dioxide forms only at pressures above 5.1 atm; the triple point of carbon dioxide is about 518 kPa at −56.6°C. The critical point is 7.38 MPa at 31.1°C. Another form of solid carbon dioxide observed at high pressure is an amorphous glass-like solid. This form of glass, called *carbonia*, is produced by supercooling heated CO_2 at extreme pressure (40–48 GPa or about 400,000 atmospheres) in a diamond anvil. This discovery confirmed the theory that carbon dioxide could exist in a glass state similar to other members of its elemental family, like silicon (silica glass) and germanium dioxide. Unlike silica and germania glasses, however, carbonia glass is not stable at normal pressures and reverts to gas when pressure is released.

At temperatures and pressures above the critical point, carbon dioxide behaves as a supercritical fluid known as supercritical carbon dioxide.

Isolation and Production

Carbon dioxide can be obtained by distillation from air, but the method is inefficient. Industrially, carbon dioxide is predominantly an unrecovered waste product, produced by several methods which may be practiced at various scales.

The combustion of all carbon-based fuels, such as methane (natural gas), petroleum distillates (gasoline, diesel, kerosene, propane), coal, wood and generic organic matter produces carbon dioxide and, except in the case of pure carbon, water. As an example, the chemical reaction between methane and oxygen is given below.

$$CH_4 + 2O_2 \rightarrow CO_2 + 2H_2O$$

It is produced by thermal decomposition of limestone, $CaCO_3$ by heating (calcining) at about 850 °C (1,560 °F), in the manufacture of quicklime (calcium oxide, CaO), a compound that has many industrial uses:

$$CaCO_3 \rightarrow CaO + CO_2$$

Iron is reduced from its oxides with coke in a blast furnace, producing pig iron and carbon dioxide:

Carbon dioxide is a byproduct of the industrial production of hydrogen by steam reforming and ammonia synthesis. These processes begin with the reaction of water and natural gas (mainly methane).

Acids liberate CO2 from most metal carbonates. Consequently, it may be obtained directly from natural carbon dioxide springs, where it is produced by the action of acidified water on limestone or dolomite. The reaction between hydrochloric acid and calcium carbonate (limestone or chalk) is shown below:

$$CaCO_3 + 2HCl \rightarrow CaCl_2 + H2CO_3$$

The carbonic acid (H_2CO_3) then decomposes to water and CO_2:

$$H_2CO_3 \rightarrow CO_2 + H_2O$$

Such reactions are accompanied by foaming or bubbling, or both, as the gas is released. They have widespread uses in industry because they can be used to neutralize waste acid streams.

Carbon dioxide is a by-product of the fermentation of sugar in the brewing of beer, whisky and other alcoholic beverages and in the production of bioethanol. Yeast metabolizes sugar to produce CO2 and ethanol, also known as alcohol, as follows:

$$C_6H_{12}O_6 \rightarrow 2CO_2 + 2C_2H_5OH$$

All aerobic organisms produce CO2 when they oxidize carbohydrates, fatty acids, and proteins. The large number of reactions involved are exceedingly complex and not described easily. Refer to (cellular respiration, anaerobic respiration and photosynthesis). The equation for the respiration of glucose and other monosaccharides is:

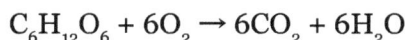

$$C_6H_{12}O_6 + 6O_2 \rightarrow 6CO_2 + 6H_2O$$

Photoautotrophs (i.e. plants and cyanobacteria) use the energy contained in sunlight to photosynthesize simple sugars from CO2 absorbed from the air and water:

$$nCO_2 + nH_2O \rightarrow (CH_2O)_n + nO_2$$

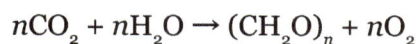

Carbon dioxide comprises about 40-45% of the gas that emanates from decomposition in landfills (termed "landfill gas"). Most of the remaining 50-55% is methane.

Uses

Carbon dioxide bubbles in a soft drink.

Carbon dioxide is used by the food industry, the oil industry, and the chemical industry. The compound has varied commercial uses but one of its greatest use as a chemical is in the production of carbonated beverages; it provides the sparkle in carbonated beverages such as soda water.

Precursor to Chemicals

In the chemical industry, carbon dioxide is mainly consumed as an ingredient in the production of urea, with a smaller fraction being used to produce methanol and a range of other products. Metal carbonates and bicarbonates, as well as some carboxylic acids derivatives (e.g., sodium salicylate) are prepared using CO_2.

In addition to conventional processes using CO_2 for chemical production, electrochemical methods are also being explored at a research level. In particular, the use of renewable energy for production of fuels from CO_2 (such as methanol) is attractive as this could result in fuels that could be easily transported and used within conventional combustion technologies but have no net CO_2 emissions.

Foods

Carbon dioxide is a food additive used as a propellant and acidity regulator in the food industry. It is approved for usage in the EU (listed as E number E290), US and Australia and New Zealand (listed by its INS number 290).

A candy called Pop Rocks is pressurized with carbon dioxide gas at about 4×10^6 Pa (40 bar, 580 psi). When placed in the mouth, it dissolves (just like other hard candy) and releases the gas bubbles with an audible pop.

Leavening agents cause dough to rise by producing carbon dioxide. Baker's yeast produces carbon dioxide by fermentation of sugars within the dough, while chemical leaveners such as baking powder and baking soda release carbon dioxide when heated or if exposed to acids.

Beverages

Carbon dioxide is used to produce carbonated soft drinks and soda water. Traditionally, the carbonation of beer and sparkling wine came about through natural fermentation, but many manufacturers carbonate these drinks with carbon dioxide recovered from the fermentation process. In the case of bottled and kegged beer, the most common method used is carbonation with recycled carbon dioxide. With the exception of British Real Ale, draught beer is usually transferred from kegs in a cold room or cellar to dispensing taps on the bar using pressurized carbon dioxide, sometimes mixed with nitrogen.

Wine Making

Carbon dioxide in the form of dry ice is often used in the wine making process to cool down clusters of grapes quickly after picking to help prevent spontaneous fermentation by wild yeast. The main advantage of using dry ice over regular water ice is that it cools the grapes without adding any additional water that may decrease the sugar concentration in the grape must, and therefore also decrease the alcohol concentration in the finished wine.

Dry ice is also used during the cold soak phase of the wine making process to keep grapes cool. The carbon dioxide gas that results from the sublimation of the dry ice tends to settle to the bottom of tanks because it is denser than air. The settled carbon dioxide gas creates a hypoxic environment which helps to prevent bacteria from growing on the grapes until it is time to start the fermentation with the desired strain of yeast.

Carbon dioxide is also used to create a hypoxic environment for carbonic maceration, the process used to produce Beaujolais wine.

Carbon dioxide is sometimes used to top up wine bottles or other storage vessels such as barrels to prevent oxidation, though it has the problem that it can dissolve into the wine, making a previously still wine slightly fizzy. For this reason, other gases such as nitrogen or argon are preferred for this process by professional wine makers.

Inert Gas

It is one of the most commonly used compressed gases for pneumatic (pressurized gas) systems in portable pressure tools. Carbon dioxide is also used as an atmosphere for welding, although in the welding arc, it reacts to oxidize most metals. Use in the automotive industry is common despite significant evidence that welds made in carbon dioxide are more brittle than those made in more inert atmospheres, and that such weld joints deteriorate over time because of the formation of carbonic acid. It is used as a welding gas primarily because it is much less expensive than more inert gases such as argon or helium. When used for MIG welding, CO_2 use is sometimes referred to as MAG welding, for Metal Active Gas, as CO_2 can react at these high temperatures. It tends to produce a hotter puddle than truly inert atmospheres, improving the flow characteristics. Although, this may be due to atmospheric reactions occurring at the puddle site. This is usually the opposite of the desired effect when welding, as it tends to embrittle the site, but may not be a problem for general mild steel welding, where ultimate ductility is not a major concern.

It is used in many consumer products that require pressurized gas because it is inexpensive and nonflammable, and because it undergoes a phase transition from gas to liquid at room temperature at an attainable pressure of approximately 60 bar (870 psi, 59 atm), allowing far more carbon dioxide to fit in a given container than otherwise would. Life jackets often contain canisters of pressured carbon dioxide for quick inflation. Aluminium capsules of CO_2 are also sold as supplies of compressed gas for airguns, paintball markers, inflating bicycle tires, and for making carbonated water. Rapid vaporization of liquid carbon dioxide is used for blasting in coal mines. High concentrations of carbon dioxide can also be used to kill pests. Liquid carbon dioxide is used in supercritical drying of some food products and technological materials, in the preparation of specimens for scanning electron microscopy and in the decaffeination of coffee beans.

Fire Extinguisher

Carbon dioxide can be used to extinguish flames by flooding the environment around the flame with the gas. It does not itself react to extinguish the flame, but starves the flame of oxygen by displacing it. Some fire extinguishers, especially those designed for electrical fires, contain liquid carbon dioxide under pressure. Carbon dioxide extinguishers work well on small flammable liquid and electrical fires, but not on ordinary combustible fires, because although it excludes oxygen, it does not cool the burning substances significantly and when the carbon dioxide disperses they are free to catch fire upon exposure to atmospheric oxygen. Their desirability in electrical fire stems from the fact that, unlike water or other chemical based methods, Carbon dioxide will not cause short circuits, leading to even more damage to equipment. Because it is a gas, it is also easy to dispense large amounts of the gas automatically in IT infrastructure rooms, where the fire itself might be hard to reach with more immediate methods because it is behind rack doors and inside of cases. Carbon dioxide has also been widely used as an extinguishing agent in fixed fire protection systems for local application of specific hazards and total flooding of a protected space. International Maritime Organization standards also recognize carbon dioxide systems for fire protection of ship holds and engine rooms. Carbon dioxide based fire protection systems have been linked to several deaths, because it can cause suffocation in sufficiently high concentrations. A review of CO_2 systems identified 51 incidents between 1975 and the date of the report, causing 72 deaths and 145 injuries.

Supercritical CO_2 as Solvent

Liquid carbon dioxide is a good solvent for many lipophilic organic compounds and is used to remove caffeine from coffee. Carbon dioxide has attracted attention in the pharmaceutical and other chemical processing industries as a less toxic alternative to more traditional solvents such as organochlorides. It is used by some dry cleaners for this reason . It is used in the preparation of some aerogels because of the properties of supercritical carbon dioxide.

Agricultural and Biological Applications

Plants require carbon dioxide to conduct photosynthesis. The atmospheres of greenhouses may (if of large size, must) be enriched with additional CO_2 to sustain and increase the rate of plant growth. At very high concentrations (100 times atmospheric concentration, or greater), carbon dioxide can be toxic to animal life, so raising the concentration to 10,000 ppm (1%) or higher for several hours will eliminate pests such as whiteflies and spider mites in a greenhouse.

In medicine, up to 5% carbon dioxide (130 times atmospheric concentration) is added to oxygen for stimulation of breathing after apnea and to stabilize the O2/CO2 balance in blood.

It has been proposed that carbon dioxide from power generation be bubbled into ponds to stimulate growth of algae that could then be converted into biodiesel fuel.

Oil Recovery

Carbon dioxide is used in enhanced oil recovery where it is injected into or adjacent to producing oil wells, usually under supercritical conditions, when it becomes miscible with the oil. This approach can increase original oil recovery by reducing residual oil saturation by between 7 per cent to 23 per cent additional to primary extraction. It acts as both a pressurizing agent and, when dissolved into the underground crude oil, significantly reduces its viscosity, and changing surface chemistry enabling the oil to flow more rapidly through the reservoir to the removal well. In mature oil fields, extensive pipe networks are used to carry the carbon dioxide to the injection points.

Bio Transformation into Fuel

Researchers have genetically modified a strain of the cyanobacterium *Synechococcus elongatus* to produce the fuels isobutyraldehyde and isobutanol from CO_2 using photosynthesis.

Refrigerant

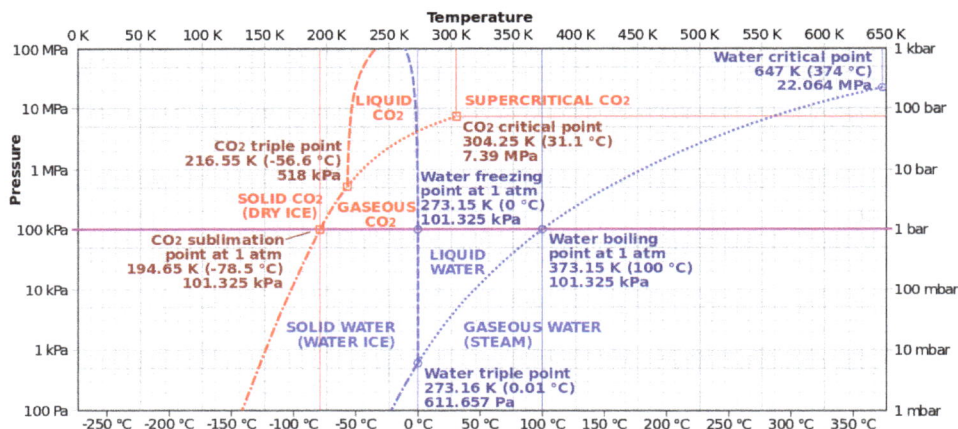

Comparison of phase diagrams of carbon dioxide (red) and water (blue) as a log-lin chart with phase transitions points at 1 atmosphere

Liquid and solid carbon dioxide are important refrigerants, especially in the food industry, where they are employed during the transportation and storage of ice cream and other frozen foods. Solid carbon dioxide is called "dry ice" and is used for small shipments where refrigeration equipment is not practical. Solid carbon dioxide is always below −78.5 °C at regular atmospheric pressure, regardless of the air temperature.

Liquid carbon dioxide (industry nomenclature R744 or R-744) was used as a refrigerant prior to the discovery of R-12 and may enjoy a renaissance due to the fact that R134a contributes to climate change. Its physical properties are highly favorable for cooling, refrigeration, and heating purposes, having a high volumetric cooling capacity. Due to the need to operate at pressures of up to 130 bar (1880 psi), CO_2 systems require highly resistant components that have already been developed for mass production in many sectors. In automobile air conditioning, in more than 90% of all driving conditions for latitudes higher than 50°, R744 operates more efficiently than systems using R134a. Its environmental advantages (GWP of 1, non-ozone depleting, non-toxic, non-flammable) could make it the future working fluid to replace current HFCs in cars, supermarkets, and heat pump water heaters, among others. Coca-Cola has fielded CO_2-based beverage coolers and the U.S. Army is interested in CO_2 refrigeration and heating technology.

The global automobile industry is expected to decide on the next-generation refrigerant in car air conditioning. CO_2 is one discussed option.

Coal bed Methane Recovery

In enhanced coal bed methane recovery, carbon dioxide would be pumped into the coal seam to displace methane, as opposed to current methods which primarily rely on the removal of water (to reduce pressure) to make the coal seam release its trapped methane.

Niche Uses

Carbon dioxide is the lasing medium in a carbon dioxide laser, which is one of the earliest type of lasers.

A carbon dioxide laser.

Carbon dioxide can be used as a means of controlling the pH of swimming pools, by continuously adding gas to the water, thus keeping the pH from rising. Among the advantages of this is the avoidance of handling (more hazardous) acids. Similarly, it is also used in the maintaining reef aquaria, where it is commonly used in calcium reactors to temporarily lower the pH of water being passed over calcium carbonate in order to allow the calcium carbonate to dissolve into the water more freely where it is used by some corals to build their skeleton.

Used as the primary coolant in the British advanced gas-cooled reactor for nuclear power generation.

Carbon dioxide induction is commonly used for the euthanasia of laboratory research animals. Methods to administer CO_2 include placing animals directly into a closed, prefilled chamber containing CO_2, or exposure to a gradually increasing concentration of CO_2. In 2013, the American Veterinary Medical Association issued new guidelines for carbon dioxide induction, stating that a displacement rate of 10% to 30% of the gas chamber volume per minute is optimal for the humane euthanization of small rodents.

Carbon dioxide is also used in several related cleaning and surface preparation techniques.

In Earth's Atmosphere

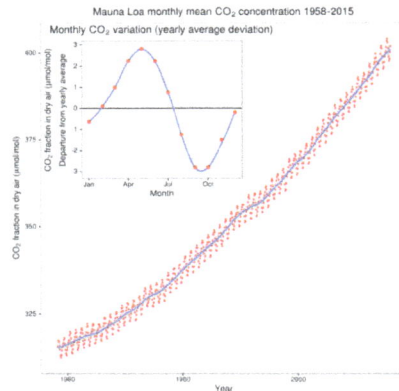

The Keeling Curve of atmospheric CO_2 concentrations measured at Mauna Loa Observatory

Carbon dioxide in Earth's atmosphere is a trace gas, currently (early 2016) having an average concentration of 402 parts per million by volume (or 611 parts per million by mass). Atmospheric concentrations of carbon dioxide fluctuate slightly with the seasons, falling during the Northern Hemisphere spring and summer as plants consume the gas and rising during northern autumn and winter as plants go dormant or die and decay. Concentrations also vary on a regional basis, most strongly near the ground with much smaller variations aloft. In urban areas concentrations are generally higher and indoors they can reach 10 times background levels.

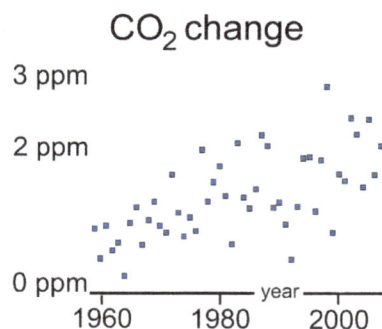

Yearly increase of atmospheric CO_2: In the 1960s, the average annual increase was 37% of the 2000–2007 average.

Combustion of fossil fuels and deforestation have caused the atmospheric concentration of carbon dioxide to increase by about 43% since the beginning of the age of industrialization. Most carbon dioxide from human activities is released from burning coal and other fossil fuels. Other human activities, including deforestation, biomass burning, and cement production also produce carbon dioxide. Volcanoes emit between 0.2 and 0.3 billion tons of carbon dioxide per year, while human activities emit about 29 billion tons.

Carbon dioxide is a greenhouse gas, absorbing and emitting infrared radiation at its two infra-red-active vibrational frequencies. This process causes carbon dioxide to warm the surface and lower atmosphere, while cooling the upper atmosphere. The increase in atmospheric concentration of CO_2, and thus in the CO_2-induced greenhouse effect, is the reason for the rise in average global temperature since the mid-20th century. Although carbon dioxide is the greenhouse gas primarily responsible for the rise, methane, nitrous oxide, ozone, and various other long-lived greenhouse gases also contribute. Carbon dioxide is of greatest concern because it exerts a larger overall warming influence than all of those other gases combined, and because it has a long atmospheric lifetime.

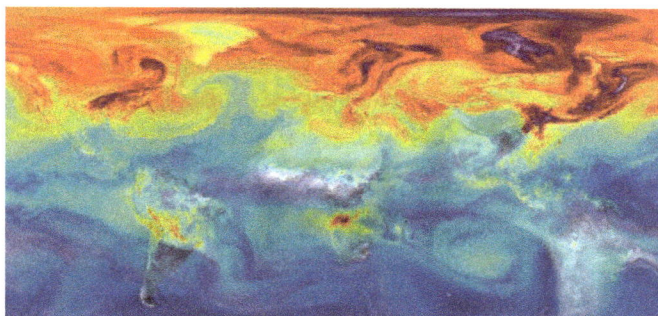

CO_2 in Earth's atmosphere if *half* of global-warming emissions are *not* absorbed.
(NASA computer simulation).

Not only do increasing carbon dioxide concentrations lead to increases in global surface tempera-ture, but increasing global temperatures also cause increasing concentrations of carbon dioxide. This produces a positive feedback for changes induced by other processes such as orbital cycles. Five hundred million years ago the carbon dioxide concentration was 20 times greater than today, decreasing to 4–5 times during the Jurassic period and then slowly declining with a particularly swift reduction occurring 49 million years ago.

Local concentrations of carbon dioxide can reach high values near strong sources, especially those that are isolated by surrounding terrain. At the Bossoleto hot spring near Rapolano Terme in Tus-cany, Italy, situated in a bowl-shaped depression about 100 m (330 ft) in diameter, concentrations of CO_2 rise to above 75% overnight, sufficient to kill insects and small animals. After sunrise the gas is dispersed by convection during the day. High concentrations of CO_2 produced by disturbance of deep lake water saturated with CO_2 are thought to have caused 37 fatalities at Lake Monoun, Cameroon in 1984 and 1700 casualties at Lake Nyos, Cameroon in 1986.

On November 12, 2015, NASA scientists reported that human-made carbon dioxide (CO_2) contin-ues to increase above levels not seen in hundreds of thousands of years: currently, about half of the carbon dioxide released from the burning of fossil fuels remains in the atmosphere and is not absorbed by vegetation and the oceans.

In the Oceans

Carbon dioxide dissolves in the ocean to form carbonic acid (H_2CO_3), bicarbonate (HCO_3^-) and carbonate (CO_3^{2-}). There is about fifty times as much carbon dissolved in the oceans as exists in the atmosphere. The oceans act as an enormous carbon sink, and have taken up about a third of CO_2 emitted by human activity.

As the concentration of carbon dioxide increases in the atmosphere, the increased uptake of carbon dioxide into the oceans is causing a measurable decrease in the pH of the oceans, which is referred to as ocean acidification. This reduction in pH affects biological systems in the oceans, primarily oceanic calcifying organisms. These effects span the food chain from autotrophs to heterotrophs and include organisms such as coccolithophores, corals, foraminifera, echinoderms, crustaceans and mollusks. Under normal conditions, calcium carbonate is stable in surface waters since the carbonate ion is at supersaturating concentrations. However, as ocean pH falls, so does the concentration of this ion, and when carbonate becomes undersaturated, structures made of calcium carbonate are vulnerable to dissolution. Corals, coccolithophore algae, coralline algae, foraminifera, shellfish and pteropods experience reduced calcification or enhanced dissolution when exposed to elevated CO2.

Gas solubility decreases as the temperature of water increases (except when both pressure exceeds 300 bar and temperature exceeds 393 K, only found near deep geothermal vents) and therefore the rate of uptake from the atmosphere decreases as ocean temperatures rise.

Most of the CO_2 taken up by the ocean, which is about 30% of the total released into the atmosphere, forms carbonic acid in equilibrium with bicarbonate. Some of these chemical species are consumed by photosynthetic organisms that remove carbon from the cycle. Increased CO_2 in the atmosphere has led to decreasing alkalinity of seawater, and there is concern that this may adversely affect organisms living in the water. In particular, with decreasing alkalinity, the availability of carbonates for forming shells decreases, although there's evidence of increased shell production by certain species under increased CO_2 content.

NOAA states in their May 2008 "State of the science fact sheet for ocean acidification" that: "The oceans have absorbed about 50% of the carbon dioxide (CO_2) released from the burning of fossil fuels, resulting in chemical reactions that lower ocean pH. This has caused an increase in hydrogen ion (acidity) of about 30% since the start of the industrial age through a process known as "ocean acidification." A growing number of studies have demonstrated adverse impacts on marine organisms, including:

- The rate at which reef-building corals produce their skeletons decreases, while production of numerous varieties of jellyfish increases.

- The ability of marine algae and free-swimming zooplankton to maintain protective shells is reduced.

- The survival of larval marine species, including commercial fish and shellfish, is reduced."

Also, the Intergovernmental Panel on Climate Change (IPCC) writes in their Climate Change 2007: Synthesis Report:

"The uptake of anthropogenic carbon since 1750 has led to the ocean becoming more acidic with an average decrease in pH of 0.1 units. Increasing atmospheric CO_2 concentrations lead to further acidification ... While the effects of observed ocean acidification on the marine biosphere are as yet undocumented, the progressive acidification of oceans is expected to have negative impacts on marine shell-forming organisms (e.g. corals) and their dependent species."

Some marine calcifying organisms (including coral reefs) have been singled out by major research agencies, including NOAA, OSPAR commission, NANOOS and the IPCC, because their most current research shows that ocean acidification should be expected to impact them negatively.

Carbon dioxide is also introduced into the oceans through hydrothermal vents. The *Champagne* hydrothermal vent, found at the Northwest Eifuku volcano at Marianas Trench Marine National Monument, produces almost pure liquid carbon dioxide, one of only two known sites in the world.

Biological Role

Carbon dioxide is an end product of cellular respiration in organisms that obtain energy by breaking down sugars, fats and amino acids with oxygen as part of their metabolism. This includes all plants, algae and animals and aerobic fungi and bacteria. In vertebrates, the carbon dioxide travels in the blood from the body's tissues to the skin (e.g., amphibians) or the gills (e.g., fish), from where it dissolves in the water, or to the lungs from where it is exhaled. During active photosynthesis, plants can absorb more carbon dioxide from the atmosphere than they release in respiration.

Photosynthesis and Carbon Fixation

Carbon fixation is a biochemical process by which atmospheric carbon dioxide is incorporated by plants, algae and (cyanobacteria) into energy-rich organic molecules such as glucose, thus creating their own food by photosynthesis. Photosynthesis uses carbon dioxide and water to produce sugars from which other organic compounds can be constructed, and oxygen is produced as a by-product.

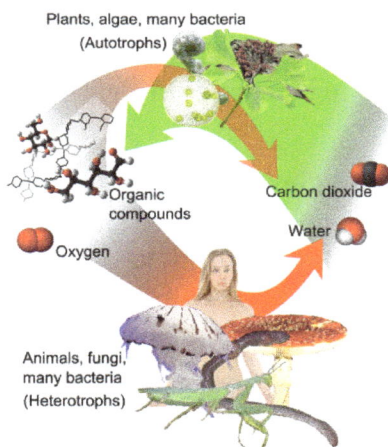

Overview of photosynthesis and respiration. Carbon dioxide (at right), together with water, form oxygen and organic compounds (at left) by photosynthesis, which can be respired to water and (CO_2).

Ribulose-1,5-bisphosphate carboxylase oxygenase, commonly abbreviated to RuBisCO, is the enzyme involved in the first major step of carbon fixation, the production of two molecules of 3-phosphoglycerate from CO_2 and ribulose bisphosphate, as shown in the diagram at left.

RuBisCO is thought to be the single most abundant protein on Earth.

Phototrophs use the products of their photosynthesis as internal food sources and as raw material for the biosynthesis of more complex organic molecules, such as polysaccharides, nucleic acids and proteins. These are used for their own growth, and also as the basis of the food chains and webs that feed other organisms, including animals such as ourselves. Some important phototrophs, the coccolithophores synthesise hard calcium carbonate scales. A globally significant species of coccolithophore is *Emiliania huxleyi* whose calcite scales have formed the basis of many sedimentary rocks such as limestone, where what was previously atmospheric carbon can remain fixed for geological timescales.

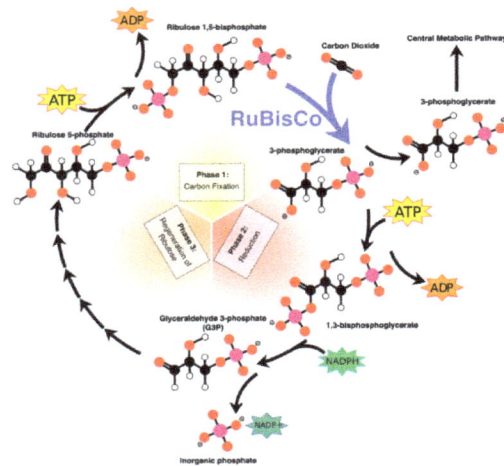

Overview of the Calvin cycle and carbon fixation

Plants can grow as much as 50 percent faster in concentrations of 1,000 ppm CO_2 when compared with ambient conditions, though this assumes no change in climate and no limitation on other nutrients. Elevated CO_2 levels cause increased growth reflected in the harvestable yield of crops, with wheat, rice and soybean all showing increases in yield of 12–14% under elevated CO_2 in FACE experiments.

Increased atmospheric CO_2 concentrations result in fewer stomata developing on plants which leads to reduced water usage and increased water-use efficiency. Studies using FACE have shown that CO_2 enrichment leads to decreased concentrations of micronutrients in crop plants. This may have knock-on effects on other parts of ecosystems as herbivores will need to eat more food to gain the same amount of protein.

The concentration of secondary metabolites such as phenylpropanoids and flavonoids can also be altered in plants exposed to high concentrations of CO_2.

Plants also emit CO_2 during respiration, and so the majority of plants and algae, which use C3 photosynthesis, are only net absorbers during the day. Though a growing forest will absorb many tons of CO_2 each year, a mature forest will produce as much CO_2 from respiration and decomposition of dead specimens (e.g., fallen branches) as is used in photosynthesis in growing plants. Contrary to the long-standing view that they are carbon neutral, mature forests can continue to accumulate carbon and remain valuable carbon sinks, helping to maintain the carbon balance of Earth's atmosphere. Additionally, and crucially to life on earth, photosynthesis by phytoplankton consumes dissolved CO_2 in the upper ocean and thereby promotes the absorption of CO_2 from the atmosphere.

Toxicity

Main symptoms of
Carbon dioxide toxicity

**Volume %
in air**

- 1%
- 3%
- 5%
- 8%

Visual —
- Dimmed
  sight

Auditory —
- Reduced
  hearing

Central
- Drowsiness
- Mild narcosis
- Dizziness
- Confusion
- Headache
- Unconsciousness

Skin
- Sweating

Respiratory —
- Shortness
  of breath

Heart
- Increased
  heart rate
  and blood
  pressure

Muscular —
- Tremor

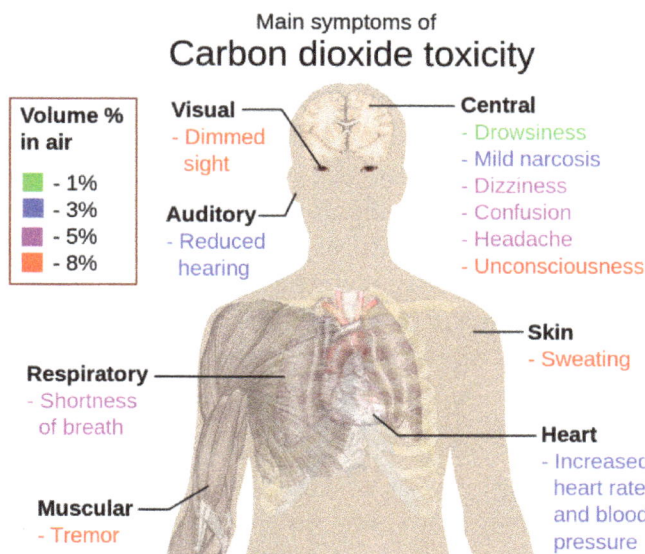

Main symptoms of carbon dioxide toxicity, by increasing volume percent in air.

Carbon dioxide content in fresh air (averaged between sea-level and 10 kPa level, i.e., about 30 km (19 mi) altitude) varies between 0.036% (360 ppm) and 0.041% (410 ppm), depending on the location.

CO_2 is an asphyxiant gas and not classified as toxic or harmful in accordance with Globally Harmonized System of Classification and Labelling of Chemicals standards of United Nations Economic Commission for Europe by using the OECD Guidelines for the Testing of Chemicals. In concentrations up to 1% (10,000 ppm), it will make some people feel drowsy and give the lungs a stuffy feeling. Concentrations of 7% to 10% (70,000 to 100,000 ppm) may cause suffocation, even in the presence of sufficient oxygen, manifesting as dizziness, headache, visual and hearing dysfunction, and unconsciousness within a few minutes to an hour. The physiological effects of acute carbon dioxide exposure are grouped together under the term hypercapnia, a subset of asphyxiation.

Because it is heavier than air, in locations where the gas seeps from the ground (due to sub-surface volcanic or geothermal activity) in relatively high concentrations, without the dispersing effects of wind, it can collect in sheltered/pocketed locations below average ground level, causing animals located therein to be suffocated. Carrion feeders attracted to the carcasses are then also killed. Children have been killed in the same way near the city of Goma by CO_2 emissions from the nearby volcano Mt. Nyiragongo. The Swahili term for this phenomenon is 'mazuku'.

Adaptation to increased concentrations of CO_2 occurs in humans, including modified breathing and kidney bicarbonate production, in order to balance the effects of blood acidification (acidosis). Several studies suggested that 2.0 percent inspired concentrations could be used for closed air spaces (e.g. a submarine) since the adaptation is physiological and reversible, as decrement in performance or in normal physical activity does not happen at this level of exposure for five days. Yet, other studies show a decrease in cognitive function even at much lower levels. Also, with ongoing respiratory acidosis, adaptation or compensatory mechanisms will be unable to reverse such condition.

Below 1%

There are few studies of the health effects of long-term continuous CO_2 exposure on humans and animals at levels below 1% and there is potentially a significant risk to humans in the near future with rising atmospheric CO_2 levels associated with climate change. Occupational CO_2 exposure limits have been set in the United States at 0.5% (5000 ppm) for an eight-hour period. At this CO_2 concentration, International Space Station crew experienced headaches, lethargy, mental slowness, emotional irritation, and sleep disruption. Studies in animals at 0.5% CO_2 have demonstrated kidney calcification and bone loss after eight weeks of exposure. A study of humans exposed in 2.5 hour sessions demonstrated significant effects on cognitive abilities at concentrations as low as 0.1% (1000ppm) CO_2 likely due to CO_2 induced increases in cerebral blood flow. Another study observed a decline in basic activity level and information usage at 1000 ppm, when compared to 500 ppm.

Ventilation

Poor ventilation is one of the main causes of excessive CO_2 concentrations in closed spaces. Carbon dioxide differential above outdoor concentrations at steady state conditions (when the occupancy and ventilation system operation are sufficiently long that CO_2 concentration has stabilized) are sometimes used to estimate ventilation rates per person. Higher CO_2 concentrations are associated with occupant health, comfort and performance degradation. ASHRAE Standard 62.1–2007 ventilation rates may result in indoor levels up to 2,100 ppm above ambient outdoor conditions. Thus if the outdoor concentration is 400 ppm, indoor concentrations may reach 2,500 ppm with ventilation rates that meet this industry consensus standard. Concentrations in poorly ventilated spaces can be found even higher than this (range of 3,000 or 4,000).

Miners, who are particularly vulnerable to gas exposure due to an insufficient ventilation, referred to mixtures of carbon dioxide and nitrogen as "blackdamp," "choke damp" or "stythe." Before more effective technologies were developed, miners would frequently monitor for dangerous levels of blackdamp and other gases in mine shafts by bringing a caged canary with them as they worked. The canary is more sensitive to asphyxiant gases than humans, and as it became unconscious would stop singing and fall off its perch. The Davy lamp could also detect high levels of blackdamp (which sinks, and collects near the floor) by burning less brightly, while methane, another suffocating gas and explosion risk, would make the lamp burn more brightly.

Human Physiology

Content

The body produces approximately 2.3 pounds (1.0 kg) of carbon dioxide per day per person, containing 0.63 pounds (290 g) of carbon. In humans, this carbon dioxide is carried through the venous system and is breathed out through the lungs, resulting in lower concentrations in the arteries. The carbon dioxide content of the blood is often given as the partial pressure, which is the pressure which carbon dioxide would have had if it alone occupied the volume.

In humans, the carbon dioxide contents are as follows:

Reference ranges or averages for partial pressures of carbon dioxide (abbreviated PCO$_2$)			
Unit	Venous blood gas	Alveolar pulmonary gas pressures	Arterial blood carbon dioxide
kPa	5.5-6.8	4.8	4.7-6.0
mmHg	41–51	36	35-45

Transport in the Blood

CO_2 is carried in blood in three different ways. (The exact percentages vary depending whether it is arterial or venous blood).

- Most of it (about 70% to 80%) is converted to bicarbonate ions HCO−3 by the enzyme carbonic anhydrase in the red blood cells, by the reaction $CO_2 + H_2O \rightarrow H_2CO_3 \rightarrow H^+ +$ HCO−3.

- 5% – 10% is dissolved in the plasma

- 5% – 10% is bound to hemoglobin as carbamino compounds

Hemoglobin, the main oxygen-carrying molecule in red blood cells, carries both oxygen and carbon dioxide. However, the CO_2 bound to hemoglobin does not bind to the same site as oxygen. Instead, it combines with the N-terminal groups on the four globin chains. However, because of allosteric effects on the hemoglobin molecule, the binding of CO_2 decreases the amount of oxygen that is bound for a given partial pressure of oxygen. The decreased binding to carbon dioxide in the blood due to increased oxygen levels is known as the Haldane Effect, and is important in the transport of carbon dioxide from the tissues to the lungs. Conversely, a rise in the partial pressure of CO_2 or a lower pH will cause offloading of oxygen from hemoglobin, which is known as the Bohr Effect.

Regulation of Respiration

Carbon dioxide is one of the mediators of local autoregulation of blood supply. If its concentration is high, the capillaries expand to allow a greater blood flow to that tissue.

Bicarbonate ions are crucial for regulating blood pH. A person's breathing rate influences the level of CO_2 in their blood. Breathing that is too slow or shallow causes respiratory acidosis, while breathing that is too rapid leads to hyperventilation, which can cause respiratory alkalosis.

Although the body requires oxygen for metabolism, low oxygen levels normally do not stimulate breathing. Rather, breathing is stimulated by higher carbon dioxide levels. As a result, breathing low-pressure air or a gas mixture with no oxygen at all (such as pure nitrogen) can lead to loss of consciousness without ever experiencing air hunger. This is especially perilous for high-altitude fighter pilots. It is also why flight attendants instruct passengers, in case of loss of cabin pressure, to apply the oxygen mask to themselves first before helping others; otherwise, one risks losing consciousness.

The respiratory centers try to maintain an arterial CO_2 pressure of 40 mm Hg. With intentional hyperventilation, the CO_2 content of arterial blood may be lowered to 10–20 mm Hg (the oxygen content of the blood is little affected), and the respiratory drive is diminished. This is why one can hold one's breath longer after hyperventilating than without hyperventilating. This carries the risk that unconsciousness may result before the need to breathe becomes overwhelming, which is why hyperventilation is particularly dangerous before free diving.

Methane

Methane is a chemical compound with the chemical formula CH_4 (one atom of carbon and four atoms of hydrogen). It is a group 14 hydride and the simplest alkane, and is the main component of natural gas. The relative abundance of methane on Earth makes it an attractive fuel, though capturing and storing it poses challenges due to its gaseous state under normal conditions for temperature and pressure.

In its natural state, methane is found both below ground and under the sea floor. When it finds its way to the surface and the atmosphere, it is known as atmospheric methane. The Earth's atmospheric methane concentration has increased by about 150% since 1750, and it accounts for 20% of the total radiative forcing from all of the long-lived and globally mixed greenhouse gases (these gases don't include water vapor which is by far the largest component of the greenhouse effect). Methane breaks down in the atmosphere and creates $CH_3 \cdot$ with water vapor.

History

In November 1776, methane was first scientifically identified by Italian physicist Alessandro Volta in the marshes of Lake Maggiore straddling Italy and Switzerland. Volta was inspired to search for the substance after reading a paper written by Benjamin Franklin about "flammable air". Volta captured the gas rising from the marsh, and by 1778 had isolated the pure gas. He also demonstrated means to ignite the gas with an electric spark.

The name "methane" was coined in 1866 by the German chemist August Wilhelm von Hofmann.

Properties and Bonding

Methane is a tetrahedral molecule with four equivalent C–H bonds. Its electronic structure is described by four bonding molecular orbitals (MOs) resulting from the overlap of the valence orbitals on C and H. The lowest energy MO is the result of the overlap of the 2s orbital on carbon with the in-phase combination of the 1s orbitals on the four hydrogen atoms. Above this energy level is a triply degenerate set of MOs that involve overlap of the 2p orbitals on carbon with various linear combinations of the 1s orbitals on hydrogen. The resulting "three-over-one" bonding scheme is consistent with photoelectron spectroscopic measurements.

At room temperature and standard pressure, methane is a colorless, odorless gas. The familiar smell of natural gas as used in homes is achieved by the addition of an odorant, usually blends containing tert-butylthiol, as a safety measure. Methane has a boiling point of −161 °C (−257.8 °F) at a pressure of one atmosphere. As a gas it is flammable over a range of concentrations (4.4–17%) in air at standard pressure.

Solid methane exists in several modifications. Presently nine are known. Cooling methane at normal pressure results in the formation of methane I. This substance crystallizes in the cubic system (space group Fm3m). The positions of the hydrogen atoms are not fixed in methane I, i.e. methane molecules may rotate freely. Therefore, it is a plastic crystal.

Chemical Reactions

The primary chemical reactions of methane are combustion, steam reforming to syngas, and halogenation. In general, methane reactions are difficult to control. Partial oxidation to methanol, for example, is challenging because the reaction typically progresses all the way to carbon dioxide and water even with an insufficient supply of oxygen. The enzyme methane monooxygenase produces methanol from methane, but cannot be used for industrial-scale reactions.

Acid-base Reactions

Like other hydrocarbons, methane is a very weak acid. Its pKa in DMSO is estimated to be 56. It cannot be deprotonated in solution, but the conjugate base with methyllithium is known.

A variety of positive ions derived from methane have been observed, mostly as unstable species in low-pressure gas mixtures. These include methenium or methyl cation CH_3^+, methane cation CH_4^+, and methanium or protonated methane CH_5^+. Some of these have been detected in outer space. Methanium can also be produced as diluted solutions from methane with superacids. Cations with higher charge, such as CH_6^{2+} and CH_7^{3+}, have been studied theoretically and conjectured to be stable.

Despite the strength of its C–H bonds, there is intense interest in catalysts that facilitate C–H bond activation in methane (and other lower numbered alkanes).

Combustion

Methane's heat of combustion is 55.5 MJ/kg. Combustion of methane is a multiple step reaction. The following equations are part of the process, with the net result being:

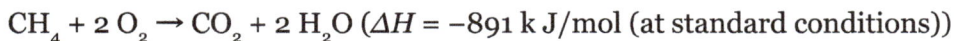

$CH_4 + 2\,O_2 \rightarrow CO_2 + 2\,H_2O$ ($\Delta H = -891$ k J/mol (at standard conditions))

1. $CH_4 + M^* \rightarrow CH_3 + H + M$

2. $CH_4 + O_2 \rightarrow CH_3 + HO_2$

3. $CH_4 + HO_2 \rightarrow CH_3 + 2\,OH$

4. $CH_4 + OH \rightarrow CH_3 + H_2O$

5. $O_2 + H \rightarrow O + OH$

6. $CH_4 + O \rightarrow CH_3 + OH$

7. $CH_3 + O_2 \rightarrow CH_2O + OH$

8. $CH_2O + O \rightarrow CHO + OH$

9. $CH_2O + OH \rightarrow CHO + H_2O$

10. $CH_2O + H \rightarrow CHO + H_2$

11. $CHO + O \rightarrow CO + OH$

12. $CHO + OH \rightarrow CO + H_2O$

13. $CHO + H \rightarrow CO + H_2$

14. $H_2 + O \rightarrow H + OH$

15. $H_2 + OH \rightarrow H + H_2O$

16. $CO + OH \rightarrow CO_2 + H$

17. $H + OH + M \rightarrow H_2O + M^*$

18. $H + H + M \rightarrow H_2 + M^*$

19. $H + O_2 + M \rightarrow HO_2 + M^*$

The species M^* signifies an energetic third body, from which energy is transferred during a molecular collision. Formaldehyde (HCHO or H2CO) is an early intermediate (reaction 7). Oxidation of formaldehyde gives the formyl radical (HCO; reactions 8–10), which then give carbon monoxide (CO) (reactions 11, 12 & 13). Any resulting H_2 oxidizes to H_2O or other intermediates (reaction 14, 15). Finally, the CO oxidizes, forming CO_2 (reaction 16). In the final stages (reactions 17–19), energy is transferred back to other third bodies. The overall speed of reaction is a function of the concentration of the various entities during the combustion process. The higher the temperature, the greater the concentration of radical species and the more rapid the combustion process.

Reactions with Halogens

Given appropriate conditions, methane reacts with halogens as follows:

$X_2 + UV \rightarrow 2\,X\bullet$

$X\bullet + CH_4 \rightarrow HX + CH_3\bullet$

$CH_3\bullet + X_2 \rightarrow CH_3X + X\bullet$

where X is a halogen: fluorine (F), chlorine (Cl), bromine (Br), or iodine (I). This mechanism for this process is called free radical halogenation. It is initiated with UV light or some other radical initiator. A chlorine atom is generated from elemental chlorine, which abstracts a hydrogen atom from methane, resulting in the formation of hydrogen chloride. The resulting methyl radical, $CH_3\bullet$, can combine with another chlorine molecule to give methyl chloride (CH_3Cl) and a chlorine atom. This chlorine atom can then react with another methane (or methyl chloride) molecule, repeating the chlorination cycle. Similar reactions can produce dichloromethane (CH_2Cl_2), chloroform ($CHCl_3$), and, ultimately, carbon tetrachloride (CCl_4), depending upon reaction conditions and the chlorine to methane ratio.

Uses

Methane is used in industrial chemical processes and may be transported as a refrigerated liquid (liquefied natural gas, or LNG). While leaks from a refrigerated liquid container are initially heavier than air due to the increased density of the cold gas, the gas at ambient temperature is lighter than air. Gas pipelines distribute large amounts of natural gas, of which methane is the principal component.

Fuel

Methane is used as a fuel for ovens, homes, water heaters, kilns, automobiles, turbines, and other things. It combusts with oxygen to create fire.

Natural Gas

Methane is important for electrical generation by burning it as a fuel in a gas turbine or steam generator. Compared to other hydrocarbon fuels, methane produces less carbon dioxide for each unit of heat released. At about 891 kJ/mol, methane's heat of combustion is lower than any other hydrocarbon but the ratio of the heat of combustion (891 kJ/mol) to the molecular mass (16.0 g/mol, of which 12.0 g/mol is carbon) shows that methane, being the simplest hydrocarbon, produces more heat per mass unit (55.7 kJ/g) than other complex hydrocarbons. In many cities, methane is piped into homes for domestic heating and cooking. In this context it is usually known as natural gas, which is considered to have an energy content of 39 megajoules per cubic meter, or 1,000 BTU per standard cubic foot.

Methane in the form of compressed natural gas is used as a vehicle fuel and is claimed to be more environmentally friendly than other fossil fuels such as gasoline/petrol and diesel. Research into adsorption methods of methane storage for use as an automotive fuel has been conducted.

Liquefied Natural Gas

Liquefied natural gas (LNG) is natural gas (predominantly methane, CH_4) that has been converted to liquid form for ease of storage or transport.

Liquefied natural gas takes up about 1/600th the volume of natural gas in the gaseous state. It is odorless, colorless, non-toxic and non-corrosive. Hazards include flammability after vaporization into a gaseous state, freezing, and asphyxia.

The liquefaction process involves removal of certain components, such as dust, acid gases, helium, water, and heavy hydrocarbons, which could cause difficulty downstream. The natural gas is then condensed into a liquid at close to atmospheric pressure (maximum transport pressure set at around 25 kPa or 3.6 psi) by cooling it to approximately –162 °C (–260 °F).

LNG achieves a higher reduction in volume than compressed natural gas (CNG) so that the energy density of LNG is 2.4 times greater than that of CNG or 60% that of diesel fuel. This makes LNG cost efficient to transport over long distances where pipelines do not exist. Specially designed cryogenic sea vessels (LNG carriers) or cryogenic road tankers are used for its transport.

LNG, when it is not highly refined for special uses, is principally used for transporting natural gas to markets, where it is regasified and distributed as pipeline natural gas. It is also beginning to be used in LNG-fueled road vehicles. For example, trucks in commercial operation have been achieving payback periods of approximately four years on the higher initial investment required in LNG equipment on the trucks and LNG infrastructure to support fueling. However, it remains more common to design vehicles to use compressed natural gas. As of 2002, the relatively higher cost of LNG production and the need to store LNG in more expensive cryogenic tanks had slowed widespread commercial use.

Liquid Methane Rocket Fuel

In a highly refined form, liquid methane is used as a rocket fuel.

Though methane has been investigated for decades, no production methane engines have yet been used on orbital spaceflights. This promises to change as liquid methane has recently been selected for the active development of a variety of bipropellant rocket engines.

Since the 1990s, a number of Russian rockets using liquid methane have been proposed. One 1990s Russian engine proposal was the RD-192, a methane/LOX variant of the RD-191.

In 2005, US companies, Orbitech and XCOR Aerospace, developed a demonstration liquid oxygen/liquid methane rocket engine and a larger 7,500 pounds-force (33 kN)-thrust engine in 2007 for potential use as the CEV lunar return engine, before the CEV program was later cancelled.

More recently the American private space company SpaceX announced in 2012 an initiative to develop liquid methane rocket engines, including initially, the very large Raptor rocket engine. Raptor is being designed to produce 4.4 meganewtons (1,000,000 lbf) of thrust with a vacuum specific impulse (I_{sp}) of 363 seconds and a sea-level I_{sp} of 321 seconds, and began component-level testing in 2014. In February 2014, the Raptor engine design was shown to be of the highly efficient and theoretically more reliable full-flow staged combustion cycle type, where both propellant streams—oxidizer and fuel—are completely in the gas phase before they enter the combustion chamber. Prior to 2014, only two full-flow rocket engines had ever progressed sufficiently to be tested on test stands, but neither engine completed development or flew on a flight vehicle.

In October 2013, the China Aerospace Science and Technology Corporation, a state-owned contractor for the Chinese space program, announced that it had completed a first ignition test on a new LOX methane rocket engine. No engine size was provided.

In September 2014, another American private space company—Blue Origin— publicly announced that they were into their third year of development work on a large methane rocket engine. The new engine, the *Blue Engine 4*, or **BE-4**, has been designed to produce 2,400 kilonewtons (550,000 lbf) of thrust. While initially planned to be used exclusively on a Blue Origin proprietary launch vehicle, it will now be used on a new United Launch Alliance (ULA) engine on an new launch vehicle that is a successor to the Atlas V. ULA indicated in 2014 that they will make the maiden flight of the new launch vehicle no earlier than 2019.

One advantage of methane is that it is abundant in many parts of the solar system and it could potentially be harvested on the surface of another solar-system body (in particular, using methane production from local materials found on Mars or Titan), providing fuel for a return journey.

By 2013, NASA's Project Morpheus had developed a small restartable LOX methane rocket engine with 5,000 pounds-force (22 kN) thrust and a specific impulse of 321 seconds suitable for inspace applications including landers. Small LOX methane thrusters 5–15 pounds-force (22–67 N) were also developed suitable for use in a Reaction Control System (RCS).

SpaceNews is reporting in early 2015 that the French space agency CNES is working with Germany and a few other governments and will propose a LOX/methane engine on a reusable launch vehicle by mid-2015, with flight testing unlikely before approximately 2026.

Chemical Feedstock

Although there is great interest in converting methane into useful or more easily liquefied compounds, the only practical processes are relatively unselective. In the chemical industry, methane is converted to synthesis gas, a mixture of carbon monoxide and hydrogen, by steam reforming. This endergonic process (requiring energy) utilizes nickel catalysts and requires high temperatures, around 700–1100 °C:

$$CH_4 + H_2O \rightarrow CO + 3\,H_2$$

Related chemistries are exploited in the Haber-Bosch Synthesis of ammonia from air, which is reduced with natural gas to a mixture of carbon dioxide, water, and ammonia.

Methane is also subjected to free-radical chlorination in the production of chloromethanes, although methanol is a more typical precursor.

Production

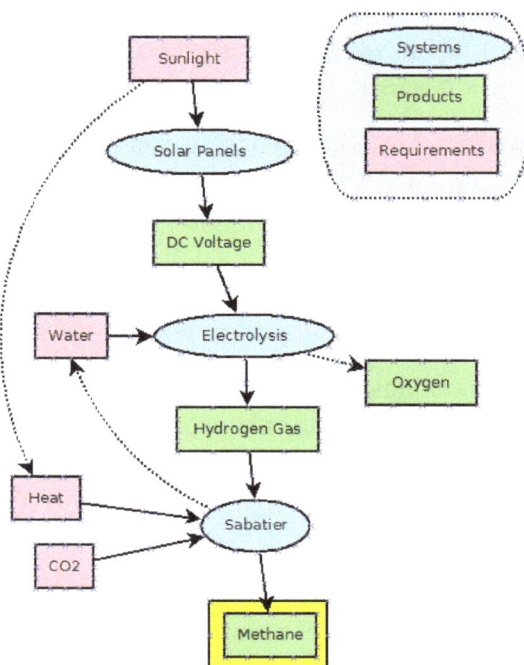

Biological Routes

Naturally occurring methane is mainly produced by the process of methanogenesis. This multistep process is used by microorganisms as an energy source. The net reaction is:

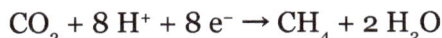

$$CO_2 + 8 H^+ + 8 e^- \rightarrow CH_4 + 2 H_2O$$

The final step in the process is catalyzed by the enzyme Coenzyme-B sulfoethylthiotransferase. Methanogenesis is a form of anaerobic respiration used by organisms that occupy landfill, ruminants (e.g., cattle), and the guts of termites.

It is uncertain if plants are a source of methane emissions.

Power to Gas

Power to gas is a technology which converts electrical power to a gas fuel. The method is used to convert carbon dioxide and water to methane, using electrolysis and the Sabatier reaction. Excess and off-peak power generated by wind generators and solar arrays could theoret-ically be used for load balancing in the energy grid.

Industrial Routes

Methane can be produced by hydrogenating carbon dioxide through the Sabatier process. Methane is also a side product of the hydrogenation of carbon monoxide in the Fischer-Tropsch process. This technology is practiced on a large scale to produce longer chain molecules than methane.

Natural gas is so abundant that the intentional production of methane is relatively rare. The only large scale facility of this kind is the Great Plains Synfuels plant, started in 1984 in Beulah, North Dakota as a way to develop abundant local resources of low grade lignite, a resource which is otherwise very hard to transport for its weight, ash content, low calorific value and propensity to spontaneous combustion during storage and transport.

An adaptation of the Sabatier methanation reaction may be used via a mixed catalyst bed and a reverse water gas shift in a single reactor to produce methane from the raw materials available on Mars, utilizing water from the Martian subsoil and carbon dioxide in the Martian atmosphere.

Laboratory Synthesis

Methane can also be produced by the destructive distillation of acetic acid in the presence of soda lime or similar. Acetic acid is decarboxylated in this process. Methane can also be prepared by re-action of aluminium carbide with water or strong acids.

Serpentinization

Methane could also be produced by a non-biological process called *serpentinization*[a] involving water, carbon dioxide, and the mineral olivine, which is known to be common on Mars.

Occurrence

Methane was discovered and isolated by Alessandro Volta between 1776 and 1778 when studying marsh gas from Lake Maggiore. It is the major component of natural gas, about 87% by volume. The major source of methane is extraction from geological deposits known as natural gas fields, with coal seam gas extraction becoming a major source (Coal bed methane extraction, a method for extracting methane from a coal deposit, while enhanced coal bed methane recovery is a method of recovering methane from non-mineable coal seams). It is associated with other hydrocarbon fuels, and sometimes accompanied by helium and nitrogen. Methane is produced at shallow levels (low pressure) by anaerobic decay of organic matter and reworked methane from deep under the Earth's surface. In general, the sediments that generate natural gas are buried deeper and at higher temperatures than those that contain oil.

Methane is generally transported in bulk by pipeline in its natural gas form, or LNG carriers in its liquefied form; few countries transport it by truck.

Alternative Sources

Testing Australian sheep for exhaled methane production (2001), CSIRO

Apart from gas fields, an alternative method of obtaining methane is via biogas generated by the fermentation of organic matter including manure, wastewater sludge, municipal solid waste (including landfills), or any other biodegradable feedstock, under anaerobic conditions. Rice fields also generate large amounts of methane during plant growth. Methane hydrates/clathrates (ice-like combinations of methane and water on the sea floor, found in vast quantities) are a potential future source of methane. Cattle belch methane accounts for 16% of the world's annual methane emissions to the atmosphere. One study reported that the livestock sector in general (primarily cattle, chickens, and pigs) produces 37% of all human-induced methane. Early research has found a number of medical treatments and dietary adjustments that help slightly limit the production of methane in ruminants. A 2009 study found that at a conservative estimate, at least 51% of global greenhouse gas emissions were attributable to the life cycle and supply chain of livestock products, meaning all meat, dairy, and by-products, and their transportation. More recently, a 2013 study estimated that livestock accounted for 44 percent of human-induced methane and 14.5 percent of human-induced greenhouse gas emissions. Many efforts are underway to reduce livestock methane production and trap the gas to use as energy.

Paleoclimatology research published in *Current Biology* suggests that flatulence from dinosaurs may have warmed the Earth.

Atmospheric Methane

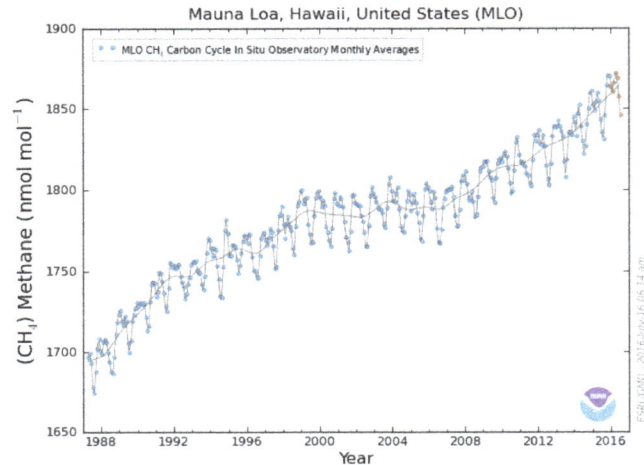

Methane concentrations up to December 2015 (Mauna Loa)

Methane is created near the Earth's surface, primarily by microorganisms by the process of methanogenesis. It is carried into the stratosphere by rising air in the tropics. Uncontrolled build-up of methane in the atmosphere is naturally checked – although human influence can upset this natural regulation – by methane's reaction with hydroxyl radicals formed from singlet oxygen atoms and with water vapor. It has a net lifetime of about 10 years, and is primarily removed by conversion to carbon dioxide and water.

In addition, there is a large (but unknown) amount of methane in methane clathrates in the ocean floors as well as the Earth's crust.

In 2010, methane levels in the Arctic were measured at 1850 nmol/mol, a level over twice as high as at any time in the 400,000 years prior to the industrial revolution. Historically, methane concentrations in the world's atmosphere have ranged between 300 and 400 nmol/mol during glacial periods commonly known as ice ages, and between 600 and 700 nmol/mol during the warm interglacial periods. Recent research suggests that the Earth's oceans are a potentially important new source of Arctic methane.

Methane is an important greenhouse gas with a global warming potential of 34 compared to CO_2 over a 100-year period, and 72 over a 20-year period.

The Earth's atmospheric methane concentration has increased by about 150% since 1750, and it accounts for 20% of the total radiative forcing from all of the long-lived and globally mixed greenhouse gases (these gases don't include water vapor which is by far the largest component of the greenhouse effect).

Clathrates

Methane is essentially insoluble in water, but it can be trapped in ice forming a similar solid. Significant deposits of methane clathrate have been found under sediments on the ocean floors of Earth at large depths.

Arctic methane release from permafrost and methane clathrates is an expected consequence and further cause of global warming.

Anaerobic Oxidation of Methane

There is a group of bacteria that drive methane oxidation with nitrite as the oxidant, the anaerobic oxidation of methane.

Safety

Methane is nontoxic, yet it is extremely flammable and may form explosive mixtures with air. Methane is violently reactive with oxidizers, halogen, and some halogen-containing compounds. Methane is also an asphyxiant and may displace oxygen in an enclosed space. Asphyxia may result if the oxygen concentration is reduced to below about 16% by displacement, as most people can tolerate a reduction from 21% to 16% without ill effects. The concentration of methane at which asphyxiation risk becomes significant is much higher than the 5–15% concentration in a flammable or explosive mixture. Methane off-gas can penetrate the interiors of buildings near landfills and expose occupants to significant levels of methane. Some buildings have specially engineered recovery systems below their basements to actively capture this gas and vent it away from the building.

Methane gas explosions are responsible for many deadly mining disasters. A methane gas explosion was the cause of the Upper Big Branch coal mine disaster in West Virginia on April 5, 2010, killing 25.

Extraterrestrial Methane

Methane has been detected or is believed to exist on all planets of the solar system and most of the larger moons. With the possible exceptions of Mars and Titan, it is believed to have come from abiotic processes.

- Mercury – the tenuous atmosphere contains trace amounts of methane.

- Venus – the atmosphere contains a large amount of methane from 60 km (37 mi) to the surface according to data collected by the Pioneer Venus Large Probe Neutral Mass Spectrometer

- Moon – traces are outgassed from the surface

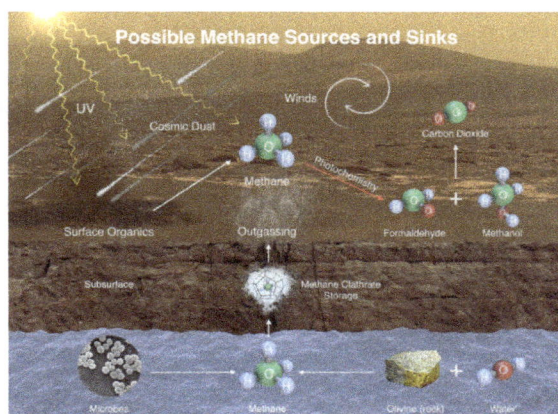

Methane (CH_4) on Mars – potential sources and sinks.

- Mars – the Martian atmosphere contains 10 nmol/mol methane. The source of methane on Mars has not been determined. Recent research suggests that methane may come from volcanoes, fault lines, or methanogens, that it may be a byproduct of electrical discharges from dust devils and dust storms, or that it may be the result of UV radiation. In January 2009, NASA scientists announced that they had discovered that the planet often vents methane into the atmosphere in specific areas, leading some to speculate this may be a sign of biological activity below the surface. Studies of a Weather Research and Forecasting model for Mars (MarsWRF) and related Mars general circulation model (MGCM) suggests that methane plume sources may be located within tens of kilometers, which is within the roving capabilities of future Mars rovers. The Curiosity rover, which landed on Mars in August 2012, can distinguish between different isotopologues of methane; but even if the mission determines that microscopic Martian life is the source of the methane, it probably resides far below the surface, beyond the rover's reach. Curiosity's Sample Analysis at Mars (SAM) instrument is capable of tracking the presence of methane over time to determine if it is constant, variable, seasonal, or random, providing further clues about its source. The first measurements with the Tunable Laser Spectrometer (TLS) indicated that there is less than 5 ppb of methane at the landing site.

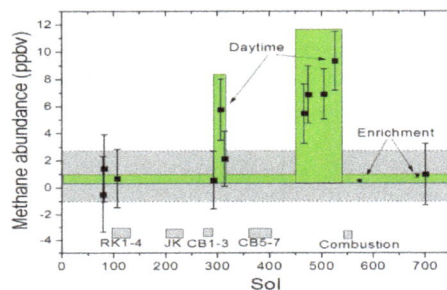

Methane measurements in the atmosphere of Mars by the *Curiosity* rover.

The Mars Trace Gas Mission orbiter planned for launch in 2016 would further study Mars' methane and its decomposition products such as formaldehyde and methanol. Alternatively, these compounds may instead be replenished by volcanic or other geological means, such as serpentinization. On July 19, 2013, NASA scientists reported finding "not much methane" (i.e., "an upper limit of 2.7 parts per billion of methane") around the Gale Crater where the Curiosity rover landed in August 2012. On September 19, 2013, from further measurements by Curiosity, NASA scientists reported no detection of atmospheric methane with a value of 0.18±0.67 ppbv corresponding to an upper limit of only 1.3 ppbv (95% confidence limit), and as a result, concluded that the probability of current methanogenic microbial activity on Mars is reduced. On 16 December 2014, NASA reported the *Curiosity* rover detected a "tenfold spike", likely localized, in the amount of methane in the Martian atmosphere. Sample measurements taken "a dozen times over 20 months" showed increases in late 2013 and early 2014, averaging "7 parts of methane per billion in the atmosphere." Before and after that, readings averaged around one-tenth that level.

- Saturn – the atmosphere contains 4500 ± 2000 ppm methane

  o Enceladus – the atmosphere contains 1.7% methane

- o Iapetus

- o Titan – the atmosphere contains 1.6% methane and thousands of methane lakes have been detected on the surface. In the upper atmosphere, methane is converted into more complex molecules including acetylene, a process that also produces molecular hydrogen. There is evidence that acetylene and hydrogen are recycled into methane near the surface. This suggests the presence either of an exotic catalyst, possibly an unknown form of methanogenic life. Methane showers, probably prompted by changing seasons, have also been observed. On October 24, 2014, methane was found in polar clouds on Titan.

Polar clouds, made of methane, on Titan (left) compared with polar clouds on Earth (right).

- Uranus – the atmosphere contains 2.3% methane

  - o Ariel – methane is believed to be a constituent of Ariel's surface ice

  - o Miranda

  - o Oberon – about 20% of Oberon's surface ice is composed of methane-related carbon/nitrogen compounds

  - o Titania – about 20% of Titania's surface ice is composed of methane-related organic compounds

  - o Umbriel – methane is a constituent of Umbriel's surface ice

- Neptune – the atmosphere contains 1.5 ± 0.5% methane

  - o Triton – Triton has a tenuous nitrogen atmosphere with small amounts of methane near the surface.

- Pluto – spectroscopic analysis of Pluto's surface reveals it to contain traces of methane

  - o Charon – methane is believed present on Charon, but it is not completely confirmed

- Eris – infrared light from the object revealed the presence of methane ice

- Halley's Comet

- Comet Hyakutake – terrestrial observations found ethane and methane in the comet

- Extrasolar planets – methane was detected on extrasolar planet HD 189733b; this is the first detection of an organic compound on a planet outside the solar system. Its origin is unknown, since the planet's high temperature (700 °C) would normally favor the formation of carbon monoxide instead. Research indicates that meteoroids slamming against exoplanet atmospheres could add hydrocarbon gases such as methane, making the exoplanets look as though they are inhabited by life, even if they are not.

- Interstellar clouds

- The atmospheres of M-type stars.

Nitrous Oxide

Nitrous oxide, commonly known as laughing gas, nitrous, nitro, or NOS is a chemical compound with the formula N2O. It is an oxide of nitrogen. At room temperature, it is a colorless, non-flammable gas, with a slightly sweet odor and taste. It is used in surgery and dentistry for its anaesthetic and analgesic effects. It is known as "laughing gas" due to the euphoric effects of inhaling it, a property that has led to its recreational use as a dissociative anaesthetic. It is also used as an oxidizer in rocket propellants, and in motor racing to increase the power output of engines. At elevated temperatures, nitrous oxide is a powerful oxidizer similar to molecular oxygen.

Nitrous oxide gives rise to nitric oxide (NO) on reaction with oxygen atoms, and this NO in turn reacts with ozone. As a result, it is the main naturally occurring regulator of stratospheric ozone. It is also a major greenhouse gas and air pollutant. Considered over a 100-year period, it is calculated to have between 265 and 310 times more impact per unit mass (global-warming potential) than carbon dioxide.

It is on the WHO Model List of Essential Medicines, the most important medications needed in a health system.

Applications

Rocket Motors

Nitrous oxide can be used as an oxidizer in a rocket motor. This has the advantages over other oxidisers in that it is not only non-toxic, but also, due to its stability at room temperature, easy to store and relatively safe to carry on a flight. As a secondary benefit it can be readily decomposed to form breathing air. Its high density and low storage pressure (when maintained at low temperature) enable it to be highly competitive with stored high-pressure gas systems.

In a 1914 patent, American rocket pioneer Robert Goddard suggested nitrous oxide and gasoline as possible propellants for a liquid-fuelled rocket. Nitrous oxide has been the oxidiser of choice in several hybrid rocket designs (using solid fuel with a liquid or gaseous oxidizer). The combination of nitrous oxide with hydroxyl-terminated polybutadiene fuel has been used by Space-ShipOne and others. It is also notably used in amateur and high power rocketry with various plastics as the fuel.

Nitrous oxide can also be used in a monopropellant rocket. In the presence of a heated catalyst, N_2O will decompose exothermically into nitrogen and oxygen, at a temperature of approximately 1,070 °F (577 °C). Because of the large heat release, the catalytic action rapidly becomes secondary as thermal autodecomposition becomes dominant. In a vacuum thruster, this can provide a monopropellant specific impulse (I_{sp}) of as much as 180 s. While noticeably less than the I_{sp} available from hydrazine thrusters (monopropellant or bipropellant with dinitrogen tetroxide), the decreased toxicity makes nitrous oxide an option worth investigating.

Nitrous oxide is said to deflagrate somewhere around 600 °C (1,112 °F) at a pressure of 21 atmospheres. At 600 psi for example, the required ignition energy is only 6 joules, whereas N_2O at 130 psi a 2500-joule ignition energy input is insufficient.

Internal Combustion Engine

In vehicle racing, nitrous oxide (often referred to as just "nitrous") allows the engine to burn more fuel by providing more oxygen than air alone, resulting in a more powerful combustion. The gas itself is not flammable at a low pressure/temperature, but it delivers more oxygen than atmospheric air by breaking down at elevated temperatures. Therefore, it is often mixed with another fuel that is easier to deflagrate. Nitrous oxide is a strong oxidant roughly equivalent to hydrogen peroxide and much stronger than oxygen gas.

Nitrous oxide is stored as a compressed liquid; the evaporation and expansion of liquid nitrous oxide in the intake manifold causes a large drop in intake charge temperature, resulting in a denser charge, further allowing more air/fuel mixture to enter the cylinder. Nitrous oxide is sometimes injected into (or prior to) the intake manifold, whereas other systems directly inject right before the cylinder (direct port injection) to increase power.

The technique was used during World War II by Luftwaffe aircraft with the GM-1 system to boost the power output of aircraft engines. Originally meant to provide the Luftwaffe standard aircraft with superior high-altitude performance, technological considerations limited its use to extremely high altitudes. Accordingly, it was only used by specialized planes like high-altitude reconnaissance aircraft, high-speed bombers, and high-altitude interceptor aircraft. It could sometimes be found on Luftwaffe aircraft also fitted with another engine-boost system, MW 50, a form of water injection for aviation engines that used methanol for its boost capabilities.

One of the major problems of using nitrous oxide in a reciprocating engine is that it can produce enough power to damage or destroy the engine. Very large power increases are possible, and if the mechanical structure of the engine is not properly reinforced, the engine may be severely damaged or destroyed during this kind of operation. It is very important with nitrous oxide augmentation of petrol engines to maintain proper operating temperatures and fuel levels to prevent "pre-ignition", or "detonation" (sometimes referred to as "knock"). Most problems that are associated with nitrous do not come from mechanical failure due to the power increases. Since nitrous allows a much denser charge into the cylinder it dramatically increases cylinder pressures. The increased pressure and temperature can cause problems such as melting the piston or valves. It may also crack or warp the piston or head and cause pre-ignition due to uneven heating.

Automotive-grade liquid nitrous oxide differs slightly from medical-grade nitrous oxide. A small amount of sulfur dioxide (SO_2) is added to prevent substance abuse. Multiple washes through a base (such as sodium hydroxide) can remove this, decreasing the corrosive properties observed when SO 2 is further oxidised during combustion into sulfuric acid, making emissions cleaner.

Aerosol Propellant

The gas is approved for use as a food additive (also known as E942), specifically as an aerosol spray propellant. Its most common uses in this context are in aerosol whipped cream canisters, cooking sprays, and as an inert gas used to displace oxygen, to inhibit bacterial growth, when filling packages of potato chips and other similar snack foods.

The gas is extremely soluble in fatty compounds. In aerosol whipped cream, it is dissolved in the fatty cream until it leaves the can, when it becomes gaseous and thus creates foam. Used in this way, it produces whipped cream four times the volume of the liquid, whereas whipping air into cream only produces twice the volume. If air were used as a propellant, oxygen would accelerate rancidification of the butterfat; nitrous oxide inhibits such degradation. Carbon dioxide cannot be used for whipped cream because it is acidic in water, which would curdle the cream and give it a seltzer-like "sparkling" sensation.

However, the whipped cream produced with nitrous oxide is unstable and will return to a more liquid state within half an hour to one hour. Thus, the method is not suitable for decorating food that will not be immediately served.

Similarly, cooking spray, which is made from various types of oils combined with lecithin (an emulsifier), may use nitrous oxide as a propellant; other propellants used in cooking spray include food-grade alcohol and propane.

Users of nitrous oxide often obtain it from whipped cream dispensers that use nitrous oxide as a propellant, for recreational use as a euphoria-inducing inhalant drug. It is not harmful in small doses, but risks due to lack of oxygen do exist.

Medicine

Nitrous oxide has been used for anaesthesia in dentistry since December 1844, where Horace Wells made the first 12–15 dental operations with the gas in Hartford. Its debut as a generally accepted method, however, came in 1863, when Gardner Quincy Colton introduced it more broadly at all the Colton Dental Association clinics, that he founded in New Haven and New York City. The first devices used in dentistry to administer the gas, known as Nitrous Oxide inhalers, were designed in a very simple way with the gas stored and breathed through a breathing bag made of rubber cloth, without a scavenger system and flowmeter, and with no addition of oxygen/air. Today these simple and somewhat unreliable inhalers have been replaced by the more modern relative analgesia machine, which is an automated machine designed to deliver a precisely dosed and breath-actuated flow of nitrous oxide mixed with oxygen, for the patient to inhale safely. The machine used in dentistry is designed as a simplified version of the larger anaesthetic machine used by hospitals, as it doesn't feature the additional anaesthetic vaporiser and medical ventilator. The purpose of the machine allows for a simpler design, as it only delivers a mixture of nitrous oxide and oxygen for the patient to inhale, in order to depress the feeling of pain while keeping the patient in a conscious state.

Medical grade N2O tanks used in dentistry.

Relative analgesia machines typically feature a constant-supply flowmeter, which allow the proportion of nitrous oxide and the combined gas flow rate to be individually adjusted. The gas is administered by dentists through a demand-valve inhaler over the nose, which will only release gas when the patient inhales through the nose. Because nitrous oxide is minimally metabolised in humans (with a rate of 0.004%), it retains its potency when exhaled into the room by the patient, and can pose an intoxicating and prolonged exposure hazard to the clinic staff if the room is poorly ventilated. Where nitrous oxide is administered, a continuous-flow fresh-air ventilation system or nitrous scavenger system is used to prevent a waste-gas buildup.

Hospitals administer nitrous oxide as one of the anaesthetic drugs delivered by anaesthetic machines. Nitrous oxide is a weak general anaesthetic, and so is generally not used alone in general anaesthesia. In general anaesthesia it is used as a carrier gas in a 2:1 ratio with oxygen for more powerful general anaesthetic drugs such as sevoflurane or desflurane. It has a minimum alveolar concentration of 105% and a blood/gas partition coefficient of 0.46.

The medical grade gas tanks, with the tradename Entonox and Nitronox contain a mixture with 50%, but this will normally be diluted to a lower percentage upon the operational delivery to the patient. Inhalation of nitrous oxide is frequently used to relieve pain associated with childbirth, trauma, oral surgery, and acute coronary syndrome (includes heart attacks). Its use during labour has been shown to be a safe and effective aid for birthing women. Its use for acute coronary syndrome is of unknown benefit.

In Britain and Canada, Entonox and Nitronox are commonly used by ambulance crews (including unregistered practitioners) as a rapid and highly effective analgesic gas.

Nitrous oxide has been shown to be effective in treating a number of addictions, including alcohol withdrawal.

Nitrous oxide is also gaining interest as a substitute gas for carbon dioxide in laparoscopic surgery. It has been found to be as safe as carbon dioxide with better pain relief.

Recreational Use

Food grade N2O whippets (above) and cracker (below)—can be used for recreational purposes

Nitrous oxide can cause analgesia, depersonalisation, derealisation, dizziness, euphoria, and some sound distortion. Research has also found that it increases suggestibility and imagination. Inhalation of nitrous oxide for recreational use, with the purpose of causing euphoria and/or slight hallucinations, began as a phenomenon for the British upper class in 1799, known as "laughing gas parties". Until at least 1863, a low availability of equipment to produce the gas, combined with a low usage of the gas for medical purposes, meant it was a relatively rare phenomenon that mainly happened among students at medical universities. When equipment became more widely available for dentistry and hospitals, most countries also restricted the legal access to buy pure nitrous oxide gas cylinders to those sectors. Despite only medical staff and dentists today being legally allowed to buy the pure gas, a Consumers Union report from 1972 found that the use of the gas for recreational purpose was [then] still taking place, based upon reports of its use in Maryland 1971, Vancouver 1972, and a survey made by Dr. Edward J. Lynn of its non-medical use in Michigan 1970.

It was not uncommon [in the interviews] to hear from individuals who had been to parties where a professional (doctor, nurse, scientist, inhalation therapist, researcher) had provided nitrous oxide. There also were those who work in restaurants who used the N2O stored in tanks for the preparation of whip cream. Reports were received from people who used the gas contained in aerosol cans both of food and non-food products. At a recent rock festival nitrous oxide was widely sold for 25 cents a balloon. Contact was made with a "mystical-religious" group that used the gas to accelerate arriving at their transcendental-meditative state of choice. Although a few, more sophisticated users employed nitrous oxide-oxygen mixes with elaborate equipment, most users used balloons or plastic bags. They either held a breath of N2O or rebreathed the gas. There were no adverse effects reported in the more than one hundred individuals surveyed.

In Australia, nitrous oxide bulbs are known as nangs, possibly derived from the sound distortion perceived by consumers.

In the United Kingdom, nitrous oxide is used by almost half a million young people at nightspots, festivals and parties. In August 2015, the London Borough of Lambeth Council banned the use of the drug for recreational purposes, making offenders liable to an on-the-spot fine of up to £1,000.

Mechanism of Action

The pharmacological mechanism of action of N2O in medicine is not fully known. However, it has been shown to directly modulate a broad range of ligand-gated ion channels, and this likely plays a major role in many of its effects. It moderately blocks NMDA and β_2-subunit-containing nACh channels, weakly inhibits AMPA, kainate, $GABA_C$, and 5-HT_3 receptors, and slightly potentiates $GABA_A$ and glycine receptors. It has also been shown to activate two-pore-domain K+ channels. While N2O affects quite a few ion channels, its anaesthetic, hallucinogenic, and euphoriant effects are likely caused predominantly or fully via inhibition of NMDA receptor-mediated currents. In addition to its effects on ion channels, N2O may act to imitate nitric oxide (NO) in the central nervous system, and this may be related to its analgesic and anxiolytic properties.

Anxiolytic Effect

In behavioural tests of anxiety, a low dose of N2O is an effective anxiolytic, and this anti-anxiety effect is associated with enhanced activity of $GABA_A$ receptors, as it is partially reversed by benzodiazepine receptor antagonists. Mirroring this, animals which have developed tolerance to the anxiolytic effects of benzodiazepines are partially tolerant to N2O. Indeed, in humans given 30% N2O, benzodiazepine receptor antagonists reduced the subjective reports of feeling "high", but did not alter psychomotor performance, in human clinical studies.

Analgesic Effect

The analgesic effects of N2O are linked to the interaction between the endogenous opioid system and the descending noradrenergic system. When animals are given morphine chronically they develop tolerance to its pain-killing effects, and this also renders the animals tolerant to the analgesic effects of N2O. Administration of antibodies which bind and block the activity of some endogenous opioids (not β-endorphin) also block the antinociceptive effects of N2O. Drugs which inhibit the breakdown of endogenous opioids also potentiate the antinociceptive effects of N2O. Several experiments have shown that opioid receptor antagonists applied directly to the brain block the antinociceptive effects of N2O, but these drugs have no effect when injected into the spinal cord.

Conversely, α_2-adrenoceptor antagonists block the pain reducing effects of N2O when given directly to the spinal cord, but not when applied directly to the brain. Indeed, α_{2B}-adrenoceptor knockout mice or animals depleted in norepinephrine are nearly completely resistant to the antinociceptive effects of N2O. Apparently N2O-induced release of endogenous opioids causes disinhibition of brain stem noradrenergic neurons, which release norepinephrine into the spinal cord and inhibit pain signalling. Exactly how N2O causes the release of endogenous opioid peptides is still uncertain.

Euphoric Effect

In rats, N2O stimulates the mesolimbic reward pathway via inducing dopamine release and activating dopaminergic neurons in the ventral tegmental area and nucleus accumbens, presumably through antagonisation of NMDA receptors localised in the system. This action has been implicated in its euphoric effects, and notably, appears to augment its analgesic properties as well.

However, it is remarkable that in mice, N2O blocks amphetamine-induced carrier-mediated dopamine release in the nucleus accumbens and behavioural sensitisation, abolishes the conditioned place preference (CPP) of cocaine and morphine, and does not produce reinforcing (or aversive) effects of its own. Studies on CPP of N2O in rats is mixed, consisting of reinforcement, aversion, and no change. In contrast, it is a positive reinforcer in squirrel monkeys, and is well known as a drug of abuse in humans. These discrepancies in response to N2O may reflect species variation or methodological differences. In human clinical studies, N2O was found to produce mixed responses similarly to rats, reflecting high subjective individual variability.

Neurotoxicity and Neuroprotection

Like other NMDA antagonists, N2O was suggested to produce neurotoxicity in the form of Olney's lesions in rodents upon prolonged (several hour) exposure. However, new research has arisen suggesting that Olney's lesions do not occur in humans, and similar drugs like ketamine are now believed not to be acutely neurotoxic. It has been argued that, because N2O has a very short duration under normal circumstances, it is less likely to be neurotoxic than other NMDA antagonists. Indeed, in rodents, short-term exposure results in only mild injury that is rapidly reversible, and permanent neuronal death only occurs after constant and sustained exposure. Nitrous oxide may also cause neurotoxicity after extended exposure because of hypoxia. This is especially true of non-medical formulations such as whipped-cream chargers (also known as "whippets" or "nangs"), which are not necessarily mixed with oxygen.

Additionally, nitrous oxide depletes vitamin B12 levels. This can cause serious neurotoxicity with even acute use if the user has preexisting vitamin B12 deficiency.

Nitrous oxide at 75-vol% reduce ischemia-induced neuronal death induced by occlusion of the middle cerebral artery in rodents, and decrease NMDA-induced Ca2+ influx in neuronal cell cultures, a critical event involved in excitotoxicity.

Safety

The major safety hazards of nitrous oxide come from the fact that it is a compressed liquefied gas, an asphyxiation risk, and a dissociative anaesthetic. Exposure to nitrous oxide causes short-term decreases in mental performance, audiovisual ability, and manual dexterity. Abusing nitrous oxide can lead to oxygen deprivation resulting in loss of blood pressure, fainting and even heart attacks.

Long-term exposure can cause vitamin B_{12} deficiency, numbness, reproductive side effects (in pregnant females), and other problems. The National Institute for Occupational Safety and Health recommends that workers' exposure to nitrous oxide should be controlled during the administration of anaesthetic gas in medical, dental, and veterinary operators. People can be exposed to nitrous oxide in the workplace by breathing it in or getting the liquid on their skin or in their eyes. The National Institute for Occupational Safety and Health (NIOSH) has set a Recommended exposure limit (REL) of 25 ppm (46 mg/m³) exposure to waste anaesthetic.

Chemical/Physical

At room temperature (20 °C (68 °F)) the saturated vapour pressure is 50.525 bar, rising up to 72.45 bar at 36.4 °C (97.5 °F)—the critical temperature. The pressure curve is thus unusually sensitive to temperature. Liquid nitrous oxide acts as a good solvent for many organic compounds; liquid mixtures may form shock sensitive explosives.

As with many strong oxidisers, contamination of parts with fuels have been implicated in rocketry accidents, where small quantities of nitrous/fuel mixtures explode due to "water hammer"-like effects (sometimes called "dieseling"—heating due to adiabatic compression of gases can reach decomposition temperatures). Some common building materials such as stainless steel and aluminium can act as fuels with strong oxidisers such as nitrous oxide, as can contaminants, which can ignite due to adiabatic compression.

There have also been accidents where nitrous oxide decomposition in plumbing has led to the explosion of large tanks.

Biological

Nitrous oxide inactivates the cobalamin form of vitamin B_{12} by oxidation. Symptoms of vitamin B_{12} deficiency, including sensory neuropathy, myelopathy, and encephalopathy, can occur within days or weeks of exposure to nitrous oxide anaesthesia in people with subclinical vitamin B_{12} deficiency. Symptoms are treated with high doses of vitamin B_{12}, but recovery can be slow and incomplete. People with normal vitamin B_{12} levels have stores to make the effects of nitrous oxide insignificant, unless exposure is repeated and prolonged (nitrous oxide abuse). Vitamin B_{12} levels should be checked in people with risk factors for vitamin B_{12} deficiency prior to using nitrous oxide anaesthesia.

A study of workers and several experimental animal studies indicate that adverse reproductive effects for pregnant females may also result from chronic exposure to nitrous oxide.

Nitrous oxide reductase is an important enzyme which limits the emission of the gas to the atmosphere.

Environmental

N2O is a greenhouse gas with a large global warming potential (GWP). When compared to carbon dioxide (CO2), N2O has 298 times the ability per molecule of gas to trap heat in the atmosphere.

N2O is produced naturally in the soil during the microbial processes of nitrification and denitrification.

The United States of America signed and ratified the United Nations Framework Convention on Climate Change (UNFCCC) in 1992, agreeing to inventory and assess the various sources of greenhouse gases that contribute to climate change. The agreement requires parties to "develop, periodically update, publish and make available... national inventories of anthropogenic emissions by sources and removals by sinks of all greenhouse gases not controlled by the Montreal Protocol, using comparable methodologies...". In response to this agreement, the U.S. is obligated to inventory anthropogenic emissions by sources and sinks, of which agriculture is a key contributor. In 2008, agriculture contributed 6.1% of the total U.S. greenhouse gas emissions and cropland contributed nearly 69% of total direct nitrous oxide (N2O) emissions. Additionally, estimated emissions from agricultural soils were 6% higher in 2008 than 1990.

According to 2006 data from the United States Environmental Protection Agency, industrial sources make up only about 20% of all anthropogenic sources, and include the production of nylon, and the burning of fossil fuel in internal combustion engines. Human activity is thought to account for 30%; tropical soils and oceanic release account for 70%. However, a 2008 study by Nobel Laureate Paul Crutzen suggests that the amount of nitrous oxide release attributable to agricultural nitrate fertilizers has been seriously underestimated, most of which would presumably come under soil and oceanic release in the Environmental Protection Agency data. Atmospheric levels have risen by more than 15% since 1750. Nitrous oxide also causes ozone depletion. A new study suggests that N_2O emission currently is the single most important ozone-depleting substance (ODS) emission and is expected to remain the largest throughout the 21st century.

Production

Nitrous oxide production

1: Ammonium nitrate
2: Bunsen burner (heating at 200° C)
3: nitrous oxide (N_2O) + water vapor
4: test tube cap
5: pipe
6: hot water (N_2O would dissolve in cold water if used)
7: sheet metal with 1/2 inch hole; holds pipe in place
8: beaker with pure N_2O

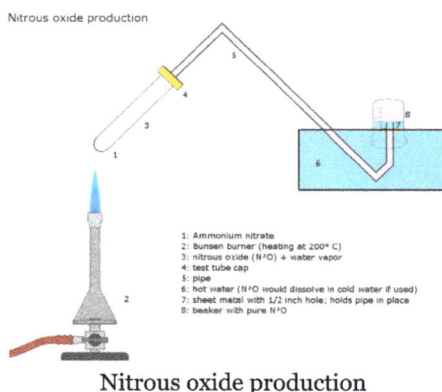

Nitrous oxide is most commonly prepared by careful heating of ammonium nitrate, which decomposes into nitrous oxide and water vapour. The addition of various phosphates favours formation of a purer gas at slightly lower temperatures. One of the earliest commercial producers was George Poe in Trenton, New Jersey.

$$NH_4NO_3 \text{ (s)} \rightarrow 2 H_2O \text{ (g)} + N_2O \text{ (g)}$$

This reaction occurs between 170 and 240 °C (338 and 464 °F), temperatures where ammonium nitrate is a moderately sensitive explosive and a very powerful oxidizer. Above 240 °C (464 °F) the exothermic reaction may accelerate to the point of detonation, so the mixture must be cooled to avoid such a disaster. Superheated steam is used to reach reaction temperature in some turnkey production plants.

Downstream, the hot, corrosive mixture of gases must be cooled to condense the steam, and filtered to remove higher oxides of nitrogen. Ammonium nitrate smoke, as an extremely persistent colloid, will also have to be removed. The cleanup is often done in a train of three gas washes; namely base, acid and base again. Any significant amounts of nitric oxide (NO) may not necessarily be absorbed directly by the base (sodium hydroxide) washes.

The nitric oxide impurity is sometimes chelated out with ferrous sulfate, reduced with iron metal, or oxidised and absorbed in base as a higher oxide. The first base wash may (or may not) react out much of the ammonium nitrate smoke. However, this reaction generates ammonia gas, which may have to be absorbed in the acid wash.

As a Byproduct

The synthesis of adipic acid; one of the two reactants used in nylon manufacture, produces nitrogen oxides including nitric oxides. This might become a major commercial source, but will require the removal of higher oxides of nitrogen and organic impurities. Currently much of the gas is decomposed before release for environmental protection.

Other Routes

Heating a mixture of sodium nitrate and ammonium sulfate.

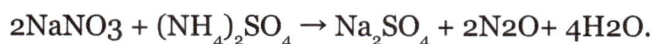

$$2NaNO_3 + (NH_4)_2SO_4 \rightarrow Na_2SO_4 + 2N_2O + 4H_2O.$$

The reaction of urea, nitric acid and sulfuric acid

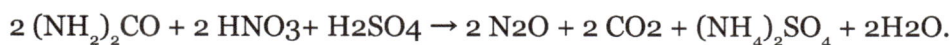

$$2 (NH_2)_2CO + 2 HNO_3 + H_2SO_4 \rightarrow 2 N_2O + 2 CO_2 + (NH_4)_2SO_4 + 2H_2O.$$

Direct oxidation of ammonia with a manganese dioxide-bismuth oxide catalyst: cf. Ostwald process.

$$2 NH_3 + 2 O_2 \rightarrow N_2O + 3 H_2O$$

Reacting Hydroxylammonium chloride with sodium nitrite. If the nitrite is added to the hydroxylamine solution, the only remaining by-product is salt water. However, if the hydroxylamine solution is added to the nitrite solution (nitrite is in excess), then toxic higher oxides of nitrogen are also formed.

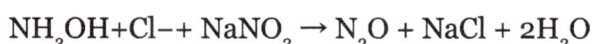

$$NH_3OH+Cl- + NaNO_2 \rightarrow N_2O + NaCl + 2H_2O$$

Reacting HNO_3 with SnCl2 and HCl:

$$2 HNO_3 + 8HCl + 4SnCl_2 \rightarrow 5H_2O + 4SnCl_4 + N_2O$$

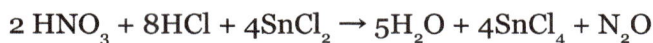

Hyponitrous acid decomposes to N_2O and water with a half-life of 16 days at 25 °C at pH 1–3.

$$H_2N_2O_2 \rightarrow H_2O + N_2O$$

Soil

Of the entire anthropogenic N2O emission (5.7 teragrams N2O-N per year), agricultural soils provide 3.5 teragrams N2O–N per year. Nitrous oxide is produced naturally in the soil during the microbial processes of nitrification, denitrification, nitrifier denitrification and others:

- aerobic autotrophic nitrification, the stepwise oxidation of ammonia (NH3) to nitrite (NO–2) and to nitrate (NO–3) (e.g., Kowalchuk and Stephen, 2001),

- anaerobic heterotrophic denitrification, the stepwise reduction of NO–3 to NO–2, nitric oxide (NO), N2O and ultimately N2, where facultative anaerobe bacteria use NO–3 as an electron acceptor in the respiration of organic material in the condition of insufficient oxygen (O 2) (e.g. Knowles, 1982), and

- nitrifier denitrification, which is carried out by autotrophic NH3–oxidizing bacteria and the pathway whereby ammonia (NH3) is oxidised to nitrite (NO–2), followed by the reduction of NO–2 to nitric oxide (NO), N2O and molecular nitrogen (N2) (e.g., Webster and Hopkins, 1996; Wrage et al., 2001).

- Other N2O production mechanisms include heterotrophic nitrification (Robertson and Kuenen, 1990), aerobic denitrification by the same heterotrophic nitrifiers (Robertson and Kuenen, 1990), fungal denitrification (Laughlin and Stevens, 2002), and non-biological process chemodenitrification (e.g. Chalk and Smith, 1983; Van Cleemput and Baert, 1984; Martikainen and De Boer, 1993; Daum and Schenk, 1998; Mørkved et al., 2007).

Soil N2O emissions are reported to be controlled by soil chemical and physical properties such as the availability of mineral N, soil pH, organic matter availability, and soil type, and climate related soil properties such as soil temperature and soil water content (e.g., Mosier, 1994; Bouwman, 1996; Beauchamp, 1997; Yamulki et al. 1997; Dobbie and Smith, 2003; Smith et al. 2003; Dalal et al. 2003).

Properties and Reactions

Nitrous oxide is a colourless, non-toxic gas with a faint, sweet odour.

Nitrous oxide supports combustion by releasing the dipolar bonded oxygen radical,thus it can relight a glowing splint.

N2O is inert at room temperature and has few reactions. At elevated temperatures, its reactivity increases. For example, nitrous oxide reacts with NaNH2 at 460 K (187 °C) to give NaN3:

$$2\ NaNH2 + N2O \rightarrow NaN3 + NaOH + NH3$$

The above reaction is the route adopted by the commercial chemical industry to produce azide salts, which are used as detonators.

Occurrence

Nitrous oxide is emitted by bacteria in soils and oceans, and is thus a part of Earth's atmosphere.

Agriculture is the main source of human-produced nitrous oxide: cultivating soil, the use of nitrogen fertilisers, and animal waste handling can all stimulate naturally occurring bacteria to produce more nitrous oxide. The livestock sector (primarily cows, chickens, and pigs) produces 65% of human-related nitrous oxide. Industrial sources make up only about 20% of all anthropogenic sources, and include the production of nylon, and the burning of fossil fuel in internal combustion engines. Human activity is thought to account for 40%; tropical soils and oceanic release account for the rest.

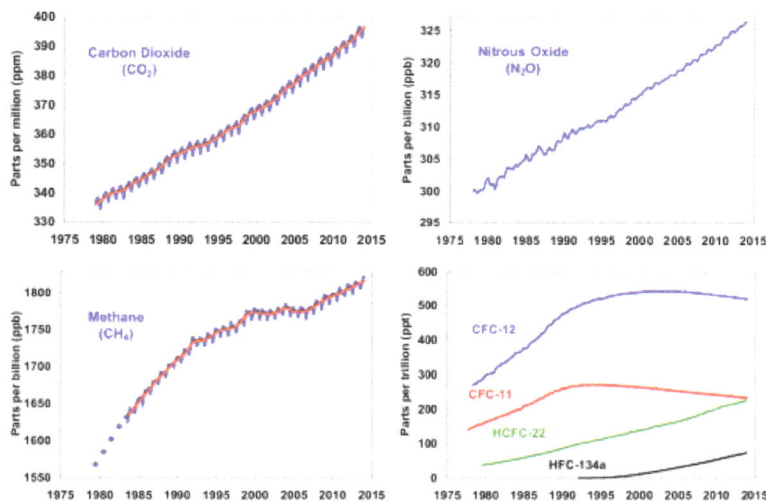

Greenhouse gas trends.

Nitrous oxide reacts with ozone in the stratosphere. Nitrous oxide is the main naturally occurring regulator of stratospheric ozone. Nitrous oxide is a major greenhouse gas. Considered over a 100-year period, it has 298 times more impact per unit weight than carbon dioxide. Thus, despite its low concentration, nitrous oxide is the fourth largest contributor to these greenhouse gases. It ranks behind water vapour, carbon dioxide, and methane. Control of nitrous oxide is part of efforts to curb greenhouse gas emissions.

History

The gas was first synthesised by English natural philosopher and chemist Joseph Priestley in 1772, who called it *phlogisticated nitrous air*. Priestley published his discovery in the book *Experiments and Observations on Different Kinds of Air (1775)*, where he described how to produce the preparation of "nitrous air diminished", by heating iron filings dampened with nitric acid.

Early use

The first important use of nitrous oxide was made possible by Thomas Beddoes and James Watt, who worked together to publish the book *Considerations on the Medical Use and on the Production of Factitious Airs (1794)*. This book was important for two reasons. First, James Watt had invented a novel machine to produce "Factitious Airs" (i.e. nitrous oxide) and a novel "breathing apparatus" to inhale the gas. Second, the book also presented the new medical theories by Thomas Beddoes, that tuberculosis and other lung diseases could be treated by inhalation of "Factitious Airs".

The machine to produce "Factitious Airs" had three parts: A furnace to burn the needed material, a vessel with water where the produced gas passed through in a spiral pipe (for impurities to be "washed off"), and finally the gas cylinder with a gasometer where the gas produced, "air", could be tapped into portable air bags (made of airtight oily silk). The breathing apparatus consisted of one of the portable air bags connected with a tube to a mouthpiece. With this new equipment being engineered and produced by 1794, the way was paved for clinical trials, which began when Thomas Beddoes in 1798 established the *"Pneumatic Institution for Relieving Diseases by Medical Airs"* in Hotwells (Bristol). In the basement of the building, a large-scale machine was producing the gases under the supervision of a young Humphry Davy, who was encouraged to experiment with new gases for patients to inhale. The first important work of Davy was examination of the nitrous oxide, and the publication of his results in the book: *Researches, Chemical and Philosophical (1800)*. In that publication, Davy notes the analgesic effect of nitrous oxide at page 465 and its potential to be used for surgical operations at page 556.

Despite Davy's discovery that inhalation of nitrous oxide could relieve a conscious person from pain, another 44 years elapsed before doctors attempted to use it for anaesthesia. The use of nitrous oxide as a recreational drug at "laughing gas parties", primarily arranged for the British upper class, became an immediate success beginning in 1799. While the effects of the gas generally make the user appear stuporous, dreamy and sedated, some people also "get the giggles" in a state of euphoria, and frequently erupt in laughter.

Anaesthetic Use

The first time nitrous oxide was used as an anaesthetic drug in the treatment of a patient was when dentist Horace Wells, with assistance by Gardner Quincy Colton and John Mankey Riggs, demonstrated insensitivity to pain from a dental extraction on 11 December 1844. In the following weeks, Wells treated the first 12–15 patients with nitrous oxide in Hartford, and according to his own record only failed in two cases. In spite of these convincing results being reported by Wells to the medical society in Boston already in December 1844, this new method was not immediately adopted by other dentists. The reason for this was most likely that Wells, in January 1845 at his first public demonstration to the medical faculty in Boston, had been partly unsuccessful, leaving his colleagues doubtful regarding its efficacy and safety. The method did not come into general use until 1863, when Gardner Quincy Colton successfully started to use it in all his "Colton Dental Association" clinics, that he had just established in New Haven and New York City. Over the following three years, Colton and his associates successfully administered nitrous oxide to more than 25,000 patients. Today, nitrous oxide is used in dentistry as an anxiolytic, as an adjunct to local anaesthetic.

However, nitrous oxide was not found to be a strong enough anaesthetic for use in major surgery in hospital settings. Being a stronger and more potent anaesthetic, sulfuric ether was instead demonstrated and accepted for use in October 1846, along with chloroform in 1847. When Joseph Thomas Clover invented the "gas-ether inhaler" in 1876, it however became a common practice at hospitals to initiate all anaesthetic treatments with a mild flow of nitrous oxide, and then gradually increase the anaesthesia with the stronger ether/chloroform. Clover's gas-ether inhaler was designed to supply the patient with nitrous oxide and ether at the same time, with the exact mixture being controlled by the operator of the device. It remained in use by many hospitals until the 1930s. Although hospitals today are using a more advanced anaesthetic machine, these machines still use the same principle launched with Clover's gas-ether inhaler, to initiate the anaesthesia with nitrous oxide, before the administration of a more powerful anaesthetic.

As a Patent Medicine

Colton's popularization of nitrous oxide led to its adoption by a number of less than reputable quacksalvers, who touted it as a cure for consumption, scrofula, catarrh, and other diseases of the blood, throat, and lungs. Nitrous oxide treatment was administered and licensed as a patent medicine by the likes of C. L. Blood and Jerome Harris in Boston and Charles E. Barney of Chicago.

Legality

In the United States, possession of nitrous oxide is legal under federal law and is not subject to DEA purview. It is, however, regulated by the Food and Drug Administration under the Food Drug and Cosmetics Act; prosecution is possible under its "misbranding" clauses, prohibiting the sale or distribution of nitrous oxide for the purpose of human consumption.

Many states have laws regulating the possession, sale, and distribution of nitrous oxide. Such laws usually ban distribution to minors or limit the amount of nitrous oxide that may be sold without special license. For example, in the state of California, possession for recreational use is prohibited and qualifies as a misdemeanour.

In New Zealand, the Ministry of Health has warned that nitrous oxide is a prescription medicine, and its sale or possession without a prescription is an offence under the Medicines Act. This statement would seemingly prohibit all non-medicinal uses of the chemical, though it is implied that only recreational use will be legally targeted.

In India, for general anaesthesia purposes, nitrous oxide is available as Nitrous Oxide IP. India's gas cylinder rules (1985) permit the transfer of gas from one cylinder to another for breathing purposes. This law benefits remote hospitals, which would otherwise suffer as a result of India's geographic immensity. Nitrous Oxide IP is transferred from bulk cylinders (17,000 litres [600 cu ft] capacity gas) to smaller pin-indexed valve cylinders (1,800 litres [64 cu ft] of gas), which are then connected to the yoke assembly of Boyle's machines. Because India's Food & Drug Authority (FDA-India) rules state that transferring a drug from one container to another (refilling) is equivalent to manufacturing, anyone found doing so must possess a drug manufacturing license.

Ozone

Ozone (systematically named $1\lambda^1,3\lambda^1$-trioxidane and catena-trioxygen), or trioxygen, is an inorganic molecule with the chemical formula O_3. It is a pale blue gas with a distinctively pungent smell. It is an allotrope of oxygen that is much less stable than the diatomic allotrope O_2, breaking down in the lower atmosphere to normal dioxygen. Ozone is formed from dioxygen by the action of ultraviolet light and also atmospheric electrical discharges, and is present in low concentrations throughout the Earth's atmosphere (stratosphere). In total, ozone makes up only 0.6 ppm of the atmosphere.

Ozone's odour is sharp, reminiscent of chlorine, and detectable by many people at concentrations of as little as 10 ppb in air. Ozone's O_3 structure was determined in 1865. The molecule was later proven to have a bent structure and to be diamagnetic. In standard conditions, ozone is a pale blue gas that condenses at progressively cryogenic temperatures to a dark blue liquid and finally a violet-black solid. Ozone's instability with regard to more common dioxygen is such that both concentrated gas and liquid ozone may decompose explosively at elevated temperatures or fast warming to the boiling point. It is therefore used commercially only in low concentrations.

Ozone is a powerful oxidant (far more so than dioxygen) and has many industrial and consumer applications related to oxidation. This same high oxidising potential, however, causes ozone to damage mucous and respiratory tissues in animals, and also tissues in plants, above concentrations of about 100 ppb. This makes ozone a potent respiratory hazard and pollutant near ground level. However, the ozone layer (a portion of the stratosphere with a bigger concentration of ozone, from two to eight ppm) is beneficial, preventing damaging ultraviolet light from reaching the Earth's surface, to the benefit of both plants and animals.

Nomenclature

The trivial name *ozone* is the most commonly used and preferred IUPAC name. The systematic names $1\lambda^1,3\lambda^1$-*trioxidane* and *catena-trioxygen*, valid IUPAC names, are constructed according to the substitutive and additive nomenclatures, respectively. The name *ozone* derives from *ozein*, the Greek verb for smell, referring to ozone's distinctive smell.

In appropriate contexts, ozone can be viewed as trioxidane with two hydrogen atoms removed, and as such, *trioxidanylidene* may be used as a context-specific systematic name, according to substitutive nomenclature. By default, these names pay no regard to the radicality of the ozone molecule. In even more specific context, this can also name the non-radical singlet ground state, whereas the diradical state is named *trioxidanediyl*.

Trioxidanediyl (or *ozonide*) is used, non-systematically, to refer to the substituent group (-OOO-). Care should be taken to avoid confusing the name of the group for the context-specific name for ozone given above.

History

In 1785, the Dutch chemist Martinus van Marum was conducting experiments involving electrical sparking above water when he noticed an unusual smell, which he attributed to the electrical reactions, failing to realize that he had in fact created ozone. A half century later, Christian Friedrich Schönbein noticed the same pungent odour and recognized it as the smell often following a bolt of lightning. In 1839, he succeeded in isolating the gaseous chemical and named it "ozone", from the Greek word *ozein* meaning "to smell". For this reason, Schönbein is generally credited with the discovery of ozone. The formula for ozone, O_3, was not determined until 1865 by Jacques-Louis Soret and confirmed by Schönbein in 1867.

Christian Friedrich Schönbein (18 October 1799 – 29 August 1868)

A prototype ozonometer built by John Smyth in 1865

For much of the second half of the nineteenth century and well into the twentieth, ozone was considered a healthy component of the environment by naturalists and health-seekers. Beaumont, California had as its official slogan "Beaumont: Zone of Ozone", as evidenced on postcards and Chamber of Commerce letterhead. Naturalists working outdoors often considered the higher elevations beneficial because of their ozone content. "There is quite a different atmosphere [at higher elevation] with enough ozone to sustain the necessary energy [to work]", wrote naturalist Henry Henshaw, working in Hawaii. Seaside air was considered to be healthy because of its believed ozone content; but the smell giving rise to this belief is in fact that of halogenated seaweed metabolites.

In fact, even Benjamin Franklin believed that the presence of cholera was connected with the deficiency or lack of ozone in the atmosphere, a sentiment shared by the British Science Association (then known simply as the British Association).

Physical Properties

Ozone is colourless or slightly bluish gas (blue when liquefied), slightly soluble in water and much more soluble in inert non-polar solvents such as carbon tetrachloride or fluorocarbons, where it forms a blue solution. At 161 K (−112 °C; −170 °F), it condenses to form a dark blue liquid. It is dangerous to allow this liquid to warm to its boiling point, because both concentrated gaseous ozone and liquid ozone can detonate. At temperatures below 80 K (−193.2 °C; −315.7 °F), it forms a violet-black solid.

Most people can detect about 0.01 μmol/mol of ozone in air where it has a very specific sharp odour somewhat resembling chlorine bleach. Exposure of 0.1 to 1 μmol/mol produces headaches, burning eyes and irritation to the respiratory passages. Even low concentrations of ozone in air are very destructive to organic materials such as latex, plastics and animal lung tissue.

Ozone is diamagnetic, which means that its electrons are all paired. In contrast, O_2 is paramagnetic, containing two unpaired electrons.

Structure

According to experimental evidence from microwave spectroscopy, ozone is a bent molecule, with C_{2v} symmetry (similar to the water molecule). The O − O distances are 127.2 pm (1.272 Å). The O − O − O angle is 116.78°. The central atom is sp^2 hybridized with one lone pair. Ozone is a polar molecule with a dipole moment of 0.53 D. The molecule can be represented as a resonance hybrid with two contributing structures, each with a single bond on one side and double bond on the other. The arrangement possesses an overall bond order of 1.5 for both sides.

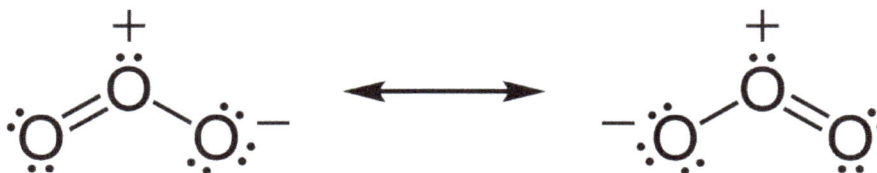

Reactions

Ozone is a powerful oxidizing agent, far stronger than O_2. It is also unstable at high concentrations, decaying to ordinary diatomic oxygen. It has a varying half-life length, depending upon atmospheric conditions (temperature, humidity, and air movement). In a sealed chamber with a fan that moves the gas, ozone has a half-life of approximately a day at room temperature. Some unverified claims imply that ozone can have a half life as short as a half an hour under atmospheric conditions.

$$2\ O3 \rightarrow 3\ O2$$

This reaction proceeds more rapidly with increasing temperature and increased pressure. Deflagration of ozone can be triggered by a spark, and can occur in ozone concentrations of 10 wt% or higher.

With Metals

Ozone will oxidise most metals (except gold, platinum, and iridium) to oxides of the metals in their highest oxidation state. For example:

$$Cu + O_3 \rightarrow CuO + O_2$$

With Nitrogen and Carbon Compounds

Ozone also oxidizes nitric oxide to nitrogen dioxide:

$$NO + O_3 \rightarrow NO_2 + O_2$$

This reaction is accompanied by chemiluminescence. The NO_2 can be further oxidized:

$$NO_2 + O_3 \rightarrow NO_3 + O_2$$

The NO_3 formed can react with NO_2 to form N_2O_5.

Solid nitronium perchlorate can be made from NO_2, ClO_2, and O_3 gases:

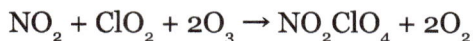

$$NO_2 + ClO_2 + 2O_3 \rightarrow NO_2ClO_4 + 2O_2$$

Ozone does not react with ammonium salts, but it oxidizes ammonia to ammonium nitrate:

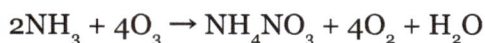

$$2NH_3 + 4O_3 \rightarrow NH_4NO_3 + 4O_2 + H_2O$$

Ozone reacts with carbon to form carbon dioxide, even at room temperature:

$$C + 2O_3 \rightarrow CO_2 + 2O_2$$

With sulfur compounds

Ozone oxidises sulfides to sulfates. For example, lead(II) sulfide is oxidised to lead(II) sulfate:

$$PbS + 4O_3 \rightarrow PbSO_4 + 4O_2$$

Sulfuric acid can be produced from ozone, water and either elemental sulfur or sulfur dioxide:

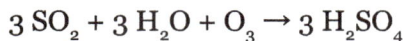

$$S + H_2O + O_3 \rightarrow H_2SO_4$$
$$3\,SO_2 + 3\,H_2O + O_3 \rightarrow 3\,H_2SO_4$$

In the gas phase, ozone reacts with hydrogen sulfide to form sulfur dioxide:

$$H_2S + O_3 \rightarrow SO_2 + H_2O$$

In an aqueous solution, however, two competing simultaneous reactions occur, one to produce elemental sulfur, and one to produce sulfuric acid:

$$H_2S + O_3 \rightarrow S + O_2 + H_2O$$
$$3\,H_2S + 4\,O_3 \rightarrow 3\,H_2SO_4$$

With Alkenes and Alkynes

Alkenes can be oxidatively cleaved by ozone, in a process called ozonolysis, giving alcohols, aldehydes, ketones, and carboxylic acids, depending on the second step of the workup.

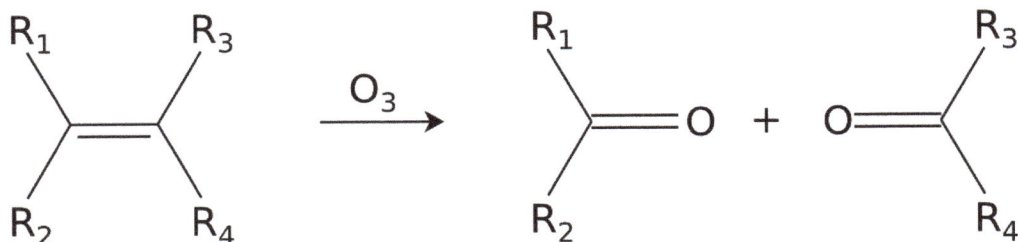

Usually ozonolysis is carried out in a solution of dichloromethane, at a temperature of −78°C. After a sequence of cleavage and rearrangement, an organic ozonide is formed. With reductive workup (e.g. zinc in acetic acid or dimethyl sulfide), ketones and aldehydes will be formed, with oxidative workup (e.g. aqueous or alcoholic hydrogen peroxide), carboxylic acids will be formed.

Other Substrates

All three atoms of ozone may also react, as in the reaction of tin(II) chloride with hydrochloric acid and ozone:

$$3 \ SnCl_2 + 6 \ HCl + O3 \rightarrow 3 \ SnCl_4 + 3 \ H_2O$$

Iodine perchlorate can be made by treating iodine dissolved in cold anhydrous perchloric acid with ozone:

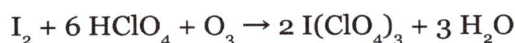

$$I_2 + 6 \ HClO_4 + O_3 \rightarrow 2 \ I(ClO_4)_3 + 3 \ H_2O$$

Combustion

Ozone can be used for combustion reactions and combustible gases; ozone provides higher temperatures than burning in dioxygen (O_2). The following is a reaction for the combustion of carbon subnitride which can also cause higher temperatures:

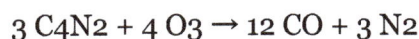

$$3 \ C4N2 + 4 \ O3 \rightarrow 12 \ CO + 3 \ N2$$

Ozone can react at cryogenic temperatures. At 77 K (−196.2 °C; −321.1 °F), atomic hydrogen reacts with liquid ozone to form a hydrogen superoxide radical, which dimerizes:

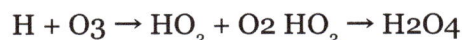

$$H + O3 \rightarrow HO_2 + O2 \ HO_2 \rightarrow H2O4$$

Reduction to Ozonides

Reduction of ozone gives the ozonide anion, O−3. Derivatives of this anion are explosive and must be stored at cryogenic temperatures. Ozonides for all the alkali metals are known. KO_3, RbO_3, and CsO_3 can be prepared from their respective superoxides:

$$KO_2 + O_3 \rightarrow KO_3 + O_2$$

Although KO_3 can be formed as above, it can also be formed from potassium hydroxide and ozone:

$$2\ KOH + 5\ O_3 \rightarrow 2\ KO_3 + 5\ O_2 + H_2O$$

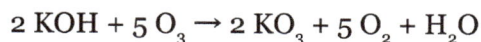

NaO_3 and LiO_3 must be prepared by action of CsO_3 in liquid NH_3 on an ion exchange resin containing Na^+ or Li^+ ions:

$$CsO_3 + Na^+ \rightarrow Cs^+ + NaO_3$$

A solution of calcium in ammonia reacts with ozone to give to ammonium ozonide and not calcium ozonide:

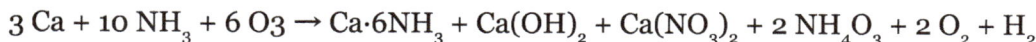

$$3\ Ca + 10\ NH_3 + 6\ O3 \rightarrow Ca \cdot 6NH_3 + Ca(OH)_2 + Ca(NO_3)_2 + 2\ NH_4O_3 + 2\ O_2 + H_2$$

Applications

Ozone can be used to remove iron and manganese from water, forming a precipitate which can be filtered:

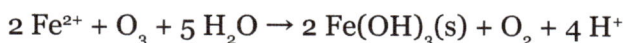

$$2\ Fe^{2+} + O_3 + 5\ H_2O \rightarrow 2\ Fe(OH)_3(s) + O_2 + 4\ H^+$$

$$2\ Mn^{2+} + 2\ O_3 + 4\ H_2O \rightarrow 2\ MnO(OH)_2(s) + 2\ O_2 + 4\ H^+$$

Ozone will also reduce dissolved hydrogen sulfide in water to sulfurous acid:

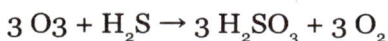

$$3\ O3 + H_2S \rightarrow 3\ H_2SO_3 + 3\ O_2$$

These three reactions are central in the use of ozone based well water treatment.

Ozone will also detoxify cyanides by converting them to cyanates.

$$CN^- + O_3 \rightarrow CNO{-} + O_2$$

Ozone will also completely decompose urea:

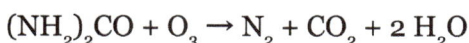

$$(NH_2)_2CO + O_3 \rightarrow N_2 + CO_2 + 2\ H_2O$$

Ozone Spectroscopy

Ozone is a bent triatomic molecule with three vibrational modes: the symmetric stretch (1103.157 cm^{-1}), bend (701.42 cm^{-1}) and antisymmetric stretch (1042.096 cm^{-1}). The symmetric stretch and bend are weak absorbers, but the antisymmetric stretch is strong and responsible for ozone being an important minor greenhouse gas. This IR band is also used to detect ambient and atmospheric ozone although UV based measurements are more common.

The electronic spectrum of ozone is quite complex. An overview can be seen at the MPI Mainz UV/VIS Spectral Atlas of Gaseous Molecules of Atmospheric Interest.

All of the bands are dissociative, meaning that the molecule falls apart to $O + O_2$ after absorbing a photon. The most important absorption is the Hartley band, extending from slightly above 300 nm down to slightly above 200 nm. It is this band that is responsible for absorbing UV C in the stratosphere.

On the high wavelength side, the Hartley band transitions to the so-called Huggins band, which falls off rapidly until disappearing by ~360 nm. Above 400 nm, extending well out into the NIR, are the Chappius and Wulf bands. There, unstructured absorption bands are useful for detecting high ambient concentrations of ozone, but are so weak that they do not have much practical effect.

There are additional absorption bands in the far UV, which increase slowly from 200 nm down to reaching a maximum at ~120 nm.

Ozone in Earth's Atmosphere

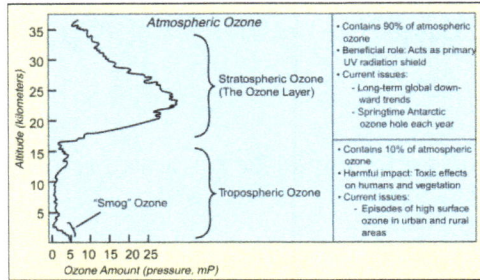

The distribution of atmospheric ozone in partial pressure as a function of altitude

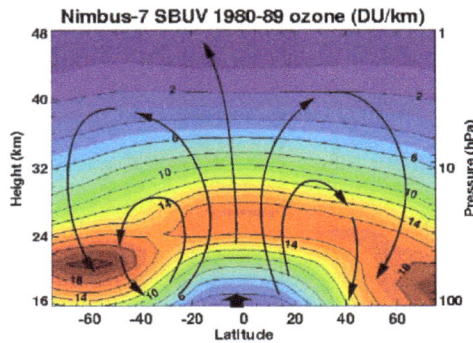

Concentration of ozone as measured by the Nimbus-7 satellite

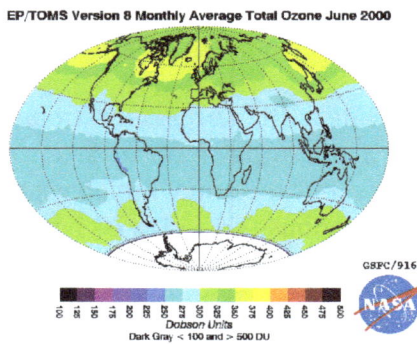

Total ozone concentration in June 2000 as measured by EP-TOMS satellite instrument

The standard way to express total ozone levels (the amount of ozone in a given vertical column) in the atmosphere is by using Dobson units. Point measurements are reported as mole fractions in nmol/mol (parts per billion, ppb) or as concentrations in $\mu g/m^3$. The study of ozone concentration in the atmosphere started in the 1920s.

Ozone Layer

Location and Production

The highest levels of ozone in the atmosphere are in the stratosphere, in a region also known as the ozone layer between about 10 km and 50 km above the surface (or between about 6 and 31 miles). However, even in this "layer", the ozone concentrations are only two to eight parts per million, so most of the oxygen there remains of the dioxygen type.

Ozone in the stratosphere is mostly produced from short-wave ultraviolet rays between 240 and 160 nm. Oxygen starts to absorb weakly at 240 nm in the Herzberg bands, but most of the oxygen is dissociated by absorption in the strong Schumann–Runge bands between 200 and 160 nm where ozone does not absorb. While shorter wavelength light, extending to even the X-Ray limit, is energetic enough to dissociate molecular oxygen, there is relatively little of it, and, the strong solar emission at Lyman-alpha, 121 nm, falls at a point where molecular oxygen absorption is a minimum.

The process of ozone creation and destruction is called the Chapman cycle and starts with the photolysis of molecular oxygen

$$O2 + photon\ (radiation\ \lambda < 240\ nm) \rightarrow 2\ O$$

followed by reaction of the oxygen atom with another molecule of oxygen to form ozone.

$$O + O2 + M \rightarrow O3 + M$$

where "M" denotes the third body that carries off the excess energy of the reaction. The ozone molecule can then absorb a UVC photon and dissociate

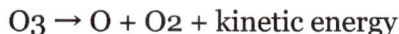

$$O3 \rightarrow O + O2 + kinetic\ energy$$

The excess kinetic energy heats the stratosphere when the O atoms and the molecular oxygen fly apart and collide with other molecules. This conversion of UV light into kinetic energy warms the stratosphere. The oxygen atoms produced in the photolysis of ozone then react back with other oxygen molecule as in the previous step to form more ozone. In the clear atmosphere, with only nitrogen and oxygen, ozone can react with the atomic oxygen to form two molecules of O_2

$$O3 + O \rightarrow 2\ O2$$

An estimate of the rate of this termination step to the cycling of atomic oxygen back to ozone can be found simply by taking the ratios of the concentration of O_2 to O_3. The termination reaction is catalysed by the presence of certain free radicals, of which the most important are hydroxyl (OH), nitric oxide (NO) and atomic chlorine (Cl) and bromine (Br). In recent decades, the amount of ozone in the stratosphere has been declining, mostly because of emissions of chlorofluorocarbons (CFC) and similar chlorinated and brominated organic molecules, which have increased the concentration of ozone-depleting catalysts above the natural background.

Importance to Surface-Dwelling Life on Earth

Ozone in the ozone layer filters out sunlight wavelengths from about 200 nm UV rays to 315 nm, with ozone peak absorption at about 250 nm. This ozone UV absorption is important to life, since it extends the absorption of UV by ordinary oxygen and nitrogen in air (which absorb all wavelengths < 200 nm) through the lower UV-C (200–280 nm) and the entire UV-B band (280–315 nm). The small unabsorbed part that remains of UV-B after passage through ozone causes sunburn in humans, and direct DNA damage in living tissues in both plants and animals. Ozone's effect on mid-range UV-B rays is illustrated by its effect on UV-B at 290 nm, which has a radiation intensity 350 million times as powerful at the top of the atmosphere as at the surface. Nevertheless, enough of UV-B radiation at similar frequency reaches the ground to cause some sunburn, and these same wavelengths are also among those responsible for the production of vitamin D in humans.

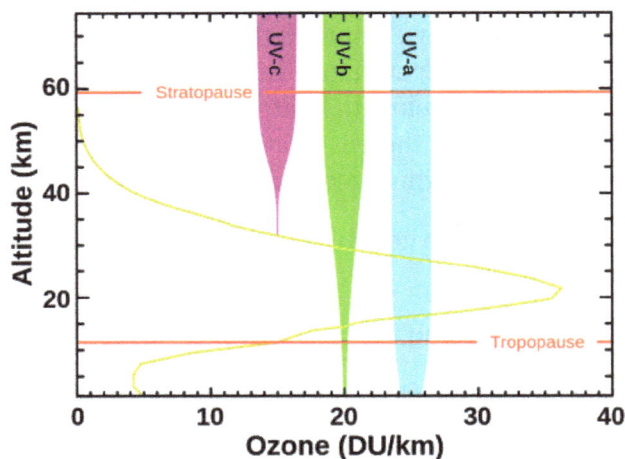

Levels of ozone at various altitudes and blocking of different bands of ultraviolet radiation. Essentially all UVC (100–280 nm) is blocked by dioxygen (at 100–200 nm) or by ozone (at 200–280 nm) in the atmosphere. The shorter portion of this band and even more energetic UV causes the formation of the ozone layer, when single oxygen atoms produced by UV photolysis of dioxygen (below 240 nm) react with more dioxygen. The ozone layer itself then blocks most, but not quite all, sunburn-producing UVB (280–315 nm). The band of UV closest to visible light, UVA (315–400 nm), is hardly affected by ozone, and most of it reaches the ground.

The ozone layer has little effect on the longer UV wavelengths called UV-A (315–400 nm), but this radiation does not cause sunburn or direct DNA damage, and while it probably does cause long-term skin damage in certain humans, it is not as dangerous to plants and to the health of surface-dwelling organisms on Earth in general.

Low Level Ozone

Low level ozone (or tropospheric ozone) is an atmospheric pollutant. It is not emitted directly by car engines or by industrial operations, but formed by the reaction of sunlight on air containing hydrocarbons and nitrogen oxides that react to form ozone directly at the source of the pollution or many kilometers down wind.

Ozone reacts directly with some hydrocarbons such as aldehydes and thus begins their removal from the air, but the products are themselves key components of smog. Ozone photolysis by UV light leads to production of the hydroxyl radical HO• and this plays a part in the removal of hydrocarbons from the air, but is also the first step in the creation of components of smog such as peroxyacyl nitrates, which can be powerful eye irritants. The atmospheric lifetime of tropospheric ozone is about 22 days; its main removal mechanisms are being deposited to the ground, the above-mentioned reaction giving HO•, and by reactions with OH and the peroxy radical HO_2•.

There is evidence of significant reduction in agricultural yields because of increased ground-level ozone and pollution which interferes with photosynthesis and stunts overall growth of some plant species. The United States Environmental Protection Agency is proposing a secondary regulation to reduce crop damage, in addition to the primary regulation designed for the protection of human health.

Certain examples of cities with elevated ozone readings are Houston, Texas, and Mexico City, Mexico. Houston has a reading of around 41 nmol/mol, while Mexico City is far more hazardous, with a reading of about 125 nmol/mol.

Ozone Cracking

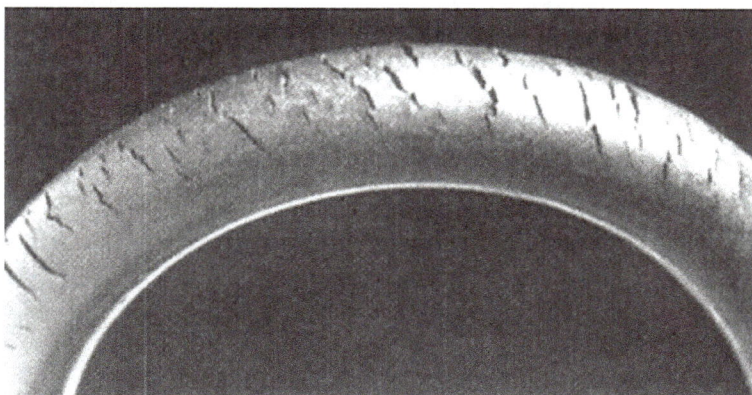

Ozone cracking in natural rubber tubing

Ozone gas attacks any polymer possessing olefinic or double bonds within its chain structure, such as natural rubber, nitrile rubber, and styrene-butadiene rubber. Products made using these polymers are especially susceptible to attack, which causes cracks to grow longer and deeper with time, the rate of crack growth depending on the load carried by the rubber component and the concentration of ozone in the atmosphere. Such materials can be protected by adding antiozonants, such as waxes, which bond to the surface to create a protective film or blend with the material and provide long term protection. Ozone cracking used to be a serious problem in car tires for example, but the problem is now seen only in very old tires. On the other hand, many critical products, like gaskets and O-rings, may be attacked by ozone produced within compressed air systems. Fuel lines made of reinforced rubber are also susceptible to attack, especially within the engine compartment, where some ozone is produced by electrical components. Storing rubber products in close proximity to a DC electric motor can accelerate ozone cracking. The commutator of the motor generates sparks which in turn produce ozone.

Ozone as a Greenhouse Gas

Although ozone was present at ground level before the Industrial Revolution, peak concentrations are now far higher than the pre-industrial levels, and even background concentrations well away from sources of pollution are substantially higher. Ozone acts as a greenhouse gas, absorbing some of the infrared energy emitted by the earth. Quantifying the greenhouse gas potency of ozone is difficult because it is not present in uniform concentrations across the globe. However, the most widely accepted scientific assessments relating to climate change (e.g. the Intergovernmental Panel on Climate Change Third Assessment Report) suggest that the radiative forcing of tropospheric ozone is about 25% that of carbon dioxide.

The annual global warming potential of tropospheric ozone is between 918–1022 tons carbon dioxide equivalent/tons tropospheric ozone. This means on a per-molecule basis, ozone in the troposphere has a radiative forcing effect roughly 1,000 times as strong as carbon dioxide. However, tropospheric ozone is a short-lived greenhouse gas, which decays in the atmosphere much more quickly than carbon dioxide. This means that over a 20-year span, the global warming potential of tropospheric ozone is much less, roughly 62 to 69 tons carbon dioxide equivalent / ton tropospheric ozone.

Because of its short-lived nature, tropospheric ozone does not have strong global effects, but has very strong radiative forcing effects on regional scales. In fact, there are regions of the world where tropospheric ozone has a radiative forcing up to 150% of carbon dioxide.

Health Effects

Ozone Air Pollution

Red Alder leaf, showing discolouration caused by ozone pollution

Ozone precursors are a group of pollutants, predominantly those emitted during the combustion of fossil fuels. Ground-level ozone pollution (tropospheric ozone) is created near the Earth's surface by the action of daylight UV rays on these precursors. The ozone at ground level is primarily from fossil fuel precursors, but methane is a natural precursor, and the very low natural background level of ozone at ground level is considered safe. This section examines the health impacts of fossil fuel burning, which raises ground level ozone far above background levels.

There is a great deal of evidence to show that ground level ozone can harm lung function and irritate the respiratory system. Exposure to ozone (and the pollutants that produce it) is linked to premature death, asthma, bronchitis, heart attack, and other cardiopulmonary problems.

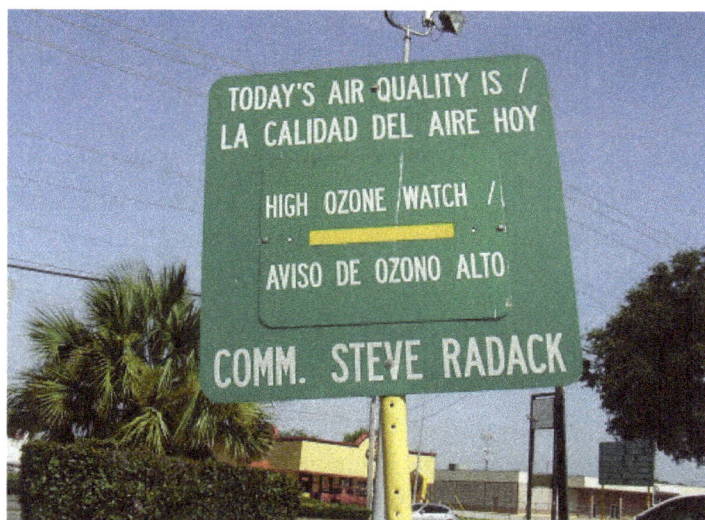

Signboard in Gulfton, Houston indicating an ozone watch

Long-term exposure to ozone has been shown to increase risk of death from respiratory illness. A study of 450,000 people living in United States cities saw a significant correlation between ozone levels and respiratory illness over the 18-year follow-up period. The study revealed that people living in cities with high ozone levels, such as Houston or Los Angeles, had an over 30% increased risk of dying from lung disease.

Air quality guidelines such as those from the World Health Organization, the United States Environmental Protection Agency (EPA) and the European Union are based on detailed studies designed to identify the levels that can cause measurable ill health effects.

According to scientists with the US EPA, susceptible people can be adversely affected by ozone levels as low as 40 nmol/mol. In the EU, the current target value for ozone concentrations is 120 µg/m^3 which is about 60 nmol/mol. This target applies to all member states in accordance with Directive 2008/50/EC. Ozone concentration is measured as a maximum daily mean of 8 hour averages and the target should not be exceeded on more than 25 calendar days per year, starting from January 2010. Whilst the directive requires in the future a strict compliance with 120 µg/m^3 limit (i.e. mean ozone concentration not to be exceeded on any day of the year), there is no date set for this requirement and this is treated as a long-term objective.

In the USA, the Clean Air Act directs the EPA to set National Ambient Air Quality Standards for several pollutants, including ground-level ozone, and counties out of compliance with these standards are required to take steps to reduce their levels. In May 2008, under a court order, the EPA lowered its ozone standard from 80 nmol/mol to 75 nmol/mol. The move proved controversial, since the Agency's own scientists and advisory board had recommended lowering the standard to 60 nmol/mol. Many public health and environmental groups also supported the 60 nmol/mol standard, and the World Health Organization recommends 51 nmol/mol.

On January 7, 2010, the U.S. Environmental Protection Agency (EPA) announced proposed revisions to the National Ambient Air Quality Standard (NAAQS) for the pollutant ozone, the principal component of smog:

... EPA proposes that the level of the 8-hour primary standard, which was set at 0.075 µmol/mol in the 2008 final rule, should instead be set at a lower level within the range of 0.060 to 0.070 µmol/mol, to provide increased protection for children and other "at risk" populations against an array of O3 - related adverse health effects that range from decreased lung function and increased respiratory symptoms to serious indicators of respiratory morbidity including emergency department visits and hospital admissions for respiratory causes, and possibly cardiovascular-related morbidity as well as total non-accidental and cardiopulmonary mortality....

On October 26, 2015, the EPA published a final rule with an effective date of December 28, 2015 that revised the 8-hour primary NAAQS from 0.075 ppm to 0.070 ppm.

The EPA has developed an Air Quality Index (AQI) to help explain air pollution levels to the general public. Under the current standards, eight-hour average ozone mole fractions of 85 to 104 nmol/mol are described as "unhealthy for sensitive groups", 105 nmol/mol to 124 nmol/mol as "unhealthy", and 125 nmol/mol to 404 nmol/mol as "very unhealthy".

Ozone can also be present in indoor air pollution, partly as a result of electronic equipment such as photocopiers. A connection has also been known to exist between the increased pollen, fungal spores, and ozone caused by thunderstorms and hospital admissions of asthma sufferers.

In the Victorian era, one British folk myth held that the smell of the sea was caused by ozone. In fact, the characteristic "smell of the sea" is caused by dimethyl sulfide, a chemical generated by phytoplankton. Victorian British folk considered the resulting smell "bracing".

Heat Waves

Ozone production rises during heat waves, because plants absorb less ozone. It is estimated that curtailed ozone absorption by plants was responsible for the loss of 460 lives in the UK in the hot summer of 2006. A similar investigation to assess the joint effects of ozone and heat during the European heat waves in 2003, concluded that these appear to be additive.

Physiology

Ozone, along with reactive forms of oxygen such as superoxide, singlet oxygen, hydrogen peroxide, and hypochlorite ions, is naturally produced by white blood cells and other biological systems (such as the roots of marigolds) as a means of destroying foreign bodies. Ozone reacts directly with organic double bonds. Also, when ozone breaks down to dioxygen it gives rise to oxygen free radicals, which are highly reactive and capable of damaging many organic molecules. Moreover, it is believed that the powerful oxidizing properties of ozone may be a contributing factor of inflammation. The cause-and-effect relationship of how the ozone is created in the body and what it does is still under consideration and still subject to various interpretations, since other body chemical processes can trigger some of the same reactions. A team headed by Paul Wentworth Jr. of the Department of Chemistry at the Scripps Research Institute has shown evidence linking the antibody-catalyzed water-oxidation pathway of the human immune response to the production of ozone. In this system, ozone is produced by antibody-catalyzed production of trioxidane from water and neutrophil-produced singlet oxygen.

When inhaled, ozone reacts with compounds lining the lungs to form specific, cholesterol-derived metabolites that are thought to facilitate the build-up and pathogenesis of atherosclerotic plaques (a form of heart disease). These metabolites have been confirmed as naturally occurring in human atherosclerotic arteries and are categorized into a class of secosterols termed *atheronals*, generated by ozonolysis of cholesterol's double bond to form a 5,6 secosterol as well as a secondary condensation product via aldolization.

Ozone has been implicated to have an adverse effect on plant growth: "... ozone reduced total chlorophylls, carotenoid and carbohydrate concentration, and increased 1-aminocyclopropane-1-carboxylic acid (ACC) content and ethylene production. In treated plants, the ascorbate leaf pool was decreased, while lipid peroxidation and solute leakage were significantly higher than in ozone-free controls. The data indicated that ozone triggered protective mechanisms against oxidative stress in citrus."

Safety Regulations

Because of the strongly oxidizing properties of ozone, ozone is a primary irritant, affecting especially the eyes and respiratory systems and can be hazardous at even low concentrations. The Canadian Center for Occupation Safety and Health reports that:

"Even very low concentrations of ozone can be harmful to the upper respiratory tract and the lungs. The severity of injury depends on both by the concentration of ozone and the duration of exposure. Severe and permanent lung injury or death could result from even a very short-term exposure to relatively low concentrations."

To protect workers potentially exposed to ozone, U.S. Occupational Safety and Health Administration has established a permissible exposure limit (PEL) of 0.1 μmol/mol (29 CFR 1910.1000 table Z-1), calculated as an 8-hour time weighted average. Higher concentrations are especially hazardous and NIOSH has established an Immediately Dangerous to Life and Health Limit (IDLH) of 5 μmol/mol. Work environments where ozone is used or where it is likely to be produced should have adequate ventilation and it is prudent to have a monitor for ozone that will alarm if the concentration exceeds the OSHA PEL. Continuous monitors for ozone are available from several suppliers.

Elevated ozone exposure can occur on passenger aircraft, with levels depending on altitude and atmospheric turbulence. United States Federal Aviation Authority regulations set a limit of 250 nmol/mol with a maximum four-hour average of 100 nmol/mol. Some planes are equipped with ozone converters in the ventilation system to reduce passenger exposure.

Production

Ozone generators are used to produce ozone for cleaning air or remove smoke odors in unoccupied rooms. These ozone generators can produce over 3 g of ozone per hour. Ozone often forms in nature under conditions where O_2 will not react. Ozone used in industry is measured in μmol/mol (ppm, parts per million), nmol/mol (ppb, parts per billion), μg/m³, mg/h (milligrams per hour) or weight percent. The regime of applied concentrations ranges from 1 to 5% in air and from 6 to 14% in oxygen for older generation methods. New electrolytic methods can achieve up 20 to 30% dissolved ozone concentrations in output water.

Ozone production demonstration, Fixed Nitrogen Research Laboratory, 1926

Temperature and humidity play a large role in how much ozone is being produced using traditional generation methods such as corona discharge and ultraviolet light. Old generation methods will produce less than 50% its nominal capacity if operated with humid ambient air than when it operates in very dry air. New generators using electrolytic methods can achieve higher purity and dissolution through using water molecules as the source of ozone production.

Corona Discharge Method

This is the most common type of ozone generator for most industrial and personal uses. While variations of the "hot spark" coronal discharge method of ozone production exist, including medical grade and industrial grade ozone generators, these units usually work by means of a corona discharge tube. They are typically cost-effective and do not require an oxygen source other than the ambient air to produce ozone concentrations of 3–6%. Fluctuations in ambient air, due to weather or other environmental conditions, cause variability in ozone production. However, they also produce nitrogen oxides as a by-product. Use of an air dryer can reduce or eliminate nitric acid formation by removing water vapor and increase ozone production. Use of an oxygen concentrator can further increase the ozone production and further reduce the risk of nitric acid formation by removing not only the water vapor, but also the bulk of the nitrogen.

Ultraviolet Light

UV ozone generators, or vacuum-ultraviolet (VUV) ozone generators, employ a light source that generates a narrow-band ultraviolet light, a subset of that produced by the Sun. The Sun's UV sustains the ozone layer in the stratosphere of Earth.

While standard UV ozone generators tend to be less expensive, they usually produce ozone with a concentration of about 0.5% or lower. Another disadvantage of this method is that it requires the air (oxygen) to be exposed to the UV source for a longer amount of time, and any gas that is not exposed to the UV source will not be treated. This makes UV generators impractical for use in situations that deal with rapidly moving air or water streams (in-duct air sterilization, for example). Production of ozone is one of the potential dangers of ultraviolet germicidal irradiation. VUV ozone generators are used in swimming pool and spa applications ranging to millions of gallons of water. VUV ozone generators, unlike corona discharge generators, do not produce harmful nitrogen by-products and also unlike corona discharge systems, VUV ozone generators work extremely well in humid air environments. There is also not normally a need for expensive off-gas mechanisms, and no need for air driers or oxygen concentrators which require extra costs and maintenance.

Cold Plasma

In the cold plasma method, pure oxygen gas is exposed to a plasma created by dielectric barrier discharge. The diatomic oxygen is split into single atoms, which then recombine in triplets to form ozone.

Cold plasma machines utilize pure oxygen as the input source and produce a maximum concentration of about 5% ozone. They produce far greater quantities of ozone in a given space of time compared to ultraviolet production. However, because cold plasma ozone generators are very expensive, they are found less frequently than the previous two types.

The discharges manifest as filamentary transfer of electrons (micro discharges) in a gap between two electrodes. In order to evenly distribute the micro discharges, a dielectric insulator must be used to separate the metallic electrodes and to prevent arcing.

Some cold plasma units also have the capability of producing short-lived allotropes of oxygen which include O_4, O_5, O_6, O_7, etc. These species are even more reactive than ordinary O3.

Electrolytic

Electrolytic ozone generation (EOG) splits water molecules into H_2, O_2, and O_3. In most EOG methods, the hydrogen gas will be removed to leave oxygen and ozone as the only reaction products. Therefore, EOG can achieve higher dissolution in water without other competing gases found in corona discharge method, such as nitrogen gases present in ambient air. This method of generation can achieve concentrations of 20–30% and is independent of air quality because water is used as the source material. Production of ozone electrolytically is typically unfavorable because of the high overpotential required to produce ozone as compared to oxygen. This is why ozone is not produced during typical water electrolysis. However, it is possible to increase the overpotential of oxygen by careful catalyst selection such that ozone is preferentially produced under electrolysis. Catalysts typically chosen for this approach are lead dioxide or boron-doped diamond.

Special Considerations

Ozone cannot be stored and transported like other industrial gases (because it quickly decays into diatomic oxygen) and must therefore be produced on site. Available ozone generators vary in the arrangement and design of the high-voltage electrodes. At production capacities higher than 20 kg per hour, a gas/water tube heat-exchanger may be utilized as ground electrode and assembled with tubular high-voltage electrodes on the gas-side. The regime of typical gas pressures is around 2 bars (200 kPa) absolute in oxygen and 3 bars (300 kPa) absolute in air. Several megawatts of electrical power may be installed in large facilities, applied as one phase AC current at 50 to 8000 Hz and peak voltages between 3,000 and 20,000 volts. Applied voltage is usually inversely related to the applied frequency.

The dominating parameter influencing ozone generation efficiency is the gas temperature, which is controlled by cooling water temperature and/or gas velocity. The cooler the water, the better the ozone synthesis. The lower the gas velocity, the higher the concentration (but the lower the net ozone produced). At typical industrial conditions, almost 90% of the effective power is dissipated as heat and needs to be removed by a sufficient cooling water flow.

Because of the high reactivity of ozone, only a few materials may be used like stainless steel (quality 316L), titanium, aluminium (as long as no moisture is present), glass, polytetrafluorethylene, or polyvinylidene fluoride. Viton may be used with the restriction of constant mechanical forces and absence of humidity (humidity limitations apply depending on the formulation). Hypalon may be used with the restriction that no water come in contact with it, except for normal atmospheric levels. Embrittlement or shrinkage is the common mode of failure of elastomers with exposure to ozone. Ozone cracking is the common mode of failure of elastomer seals like O-rings.

Silicone rubbers are usually adequate for use as gaskets in ozone concentrations below 1 wt%, such as in equipment for accelerated aging of rubber samples.

Incidental Production

Ozone may be formed from O2 by electrical discharges and by action of high energy electromagnetic radiation. Unsuppressed arcing in electrical contacts, motor bushes, or mechanical switches breaks down the chemical bonds of the atmospheric oxygen surrounding the contacts [O2 → 2O]. Free radicals of oxygen in and around the arc recombine to create ozone [O3]. Certain electrical equipment generate significant levels of ozone. This is especially true of devices using high voltages, such as ionic air purifiers, laser printers, photocopiers, tasers and arc welders. Electric motors using brushes can generate ozone from repeated sparking inside the unit. Large motors that use brushes, such as those used by elevators or hydraulic pumps, will generate more ozone than smaller motors.

Ozone is similarly formed in the Catatumbo lightning storms phenomenon on the Catatumbo River in Venezuela, which helps to replenish ozone in the upper troposphere. It is the world's largest single natural generator of ozone, lending calls for it to be designated a UNESCO World Heritage Site.

Laboratory Production

In the laboratory, ozone can be produced by electrolysis using a 9 volt battery, a pencil graphite rod cathode, a platinum wire anode and a 3 molar sulfuric acid electrolyte. The half cell reactions taking place are:

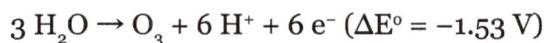

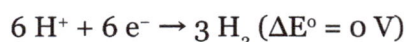

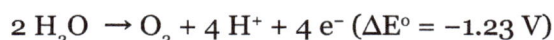

$$3\ H_2O \rightarrow O_3 + 6\ H^+ + 6\ e^-\ (\Delta E^\circ = -1.53\ V)$$

$$6\ H^+ + 6\ e^- \rightarrow 3\ H_2\ (\Delta E^\circ = 0\ V)$$

$$2\ H_2O \rightarrow O_2 + 4\ H^+ + 4\ e^-\ (\Delta E^\circ = -1.23\ V)$$

In the net reaction, three equivalents of water are converted into one equivalent of ozone and three equivalents of hydrogen. Oxygen formation is a competing reaction.

It can also be generated by a high voltage arc. In its simplest form, high voltage AC, such as the output of a Neon-sign transformer is connected to two metal rods with the ends placed sufficiently close to each other to allow an arc. The resulting arc will convert atmospheric oxygen to ozone.

It is often desirable to contain the ozone. This can be done with an apparatus consisting of two concentric glass tubes sealed together at the top with gas ports at the top and bottom of the outer tube. The inner core should have a length of metal foil inserted into it connected to one side of the power source. The other side of the power source should be connected to another piece of foil wrapped around the outer tube. A source of dry O2 is applied to the bottom port. When high voltage is applied to the foil leads, electricity will discharge between the dry dioxygen in the middle and form O3 and O2 which will flow out the top port. The reaction can be summarized as follows:

$$3\ O_2 - electricity \rightarrow 2\ O_3$$

Applications

Industry

The largest use of ozone is in the preparation of pharmaceuticals, synthetic lubricants, and many other commercially useful organic compounds, where it is used to sever carbon-carbon bonds. It can also be used for bleaching substances and for killing microorganisms in air and water sources. Many municipal drinking water systems kill bacteria with ozone instead of the more common chlorine. Ozone has a very high oxidation potential. Ozone does not form organochlorine compounds, nor does it remain in the water after treatment. Ozone can form the suspected carcinogen bromate in source water with high bromide concentrations. The U.S. Safe Drinking Water Act mandates that these systems introduce an amount of chlorine to maintain a minimum of 0.2 μmol/mol residual free chlorine in the pipes, based on results of regular testing. Where electrical power is abundant, ozone is a cost-effective method of treating water, since it is produced on demand and does not require transportation and storage of hazardous chemicals. Once it has decayed, it leaves no taste or odour in drinking water.

Although low levels of ozone have been advertised to be of some disinfectant use in residential homes, the concentration of ozone in dry air required to have a rapid, substantial effect on airborne pathogens exceeds safe levels recommended by the U.S. Occupational Safety and Health Administration and Environmental Protection Agency. Humidity control can vastly improve both the killing power of the ozone and the rate at which it decays back to oxygen (more humidity allows more effectiveness). Spore forms of most pathogens are very tolerant of atmospheric ozone in concentrations where asthma patients start to have issues.

Industrially, ozone is used to:

- Disinfect laundry in hospitals, food factories, care homes etc.;
- Disinfect water in place of chlorine

- Deodorize air and objects, such as after a fire. This process is extensively used in fabric restoration

- Kill bacteria on food or on contact surfaces;

- Sanitize swimming pools and spas

- Kill insects in stored grain

- Scrub yeast and mold spores from the air in food processing plants;

- Wash fresh fruits and vegetables to kill yeast, mold and bacteria;

- Chemically attack contaminants in water (iron, arsenic, hydrogen sulfide, nitrites, and complex organics lumped together as "colour");

- Provide an aid to flocculation (agglomeration of molecules, which aids in filtration, where the iron and arsenic are removed);

- Manufacture chemical compounds via chemical synthesis

- Clean and bleach fabrics (the former use is utilized in fabric restoration; the latter use is patented);

- Act as an antichlor in chlorine-based bleaching;

- Assist in processing plastics to allow adhesion of inks;

- Age rubber samples to determine the useful life of a batch of rubber;

- Eradicate water borne parasites such as *Giardia lamblia* and *Cryptosporidium* in surface water treatment plants.

Ozone is a reagent in many organic reactions in the laboratory and in industry. Ozonolysis is the cleavage of an alkene to carbonyl compounds.

Many hospitals around the world use large ozone generators to decontaminate operating rooms between surgeries. The rooms are cleaned and then sealed airtight before being filled with ozone which effectively kills or neutralizes all remaining bacteria.

Ozone is used as an alternative to chlorine or chlorine dioxide in the bleaching of wood pulp. It is often used in conjunction with oxygen and hydrogen peroxide to eliminate the need for chlorine-containing compounds in the manufacture of high-quality, white paper.

Ozone can be used to detoxify cyanide wastes (for example from gold and silver mining) by oxidising cyanide to cyanate and eventually to carbon dioxide.

Consumers

Devices generating high levels of ozone, some of which use ionization, are used to sanitize and deodorize uninhabited buildings, rooms, ductwork, woodsheds, boats and other vehicles.

In the U.S., air purifiers emitting low levels of ozone have been sold. This kind of air purifier is sometimes claimed to imitate nature's way of purifying the air without filters and to sanitize both it and household surfaces. The United States Environmental Protection Agency (EPA) has declared that there is "evidence to show that at concentrations that do not exceed public health standards, ozone is not effective at removing many odor-causing chemicals" or "viruses, bacteria, mold, or other biological pollutants". Furthermore, its report states that "results of some controlled studies show that concentrations of ozone considerably higher than these [human safety] standards are possible even when a user follows the manufacturer's operating instructions". A couple kept repeating health claims for the generator they sold, without supporting scientific studies. In 1998, a federal jury convicted them, among others things, of illegally distributing an ozone generator and of wire fraud.

Ozonated water is used to launder clothes and to sanitize food, drinking water, and surfaces in the home. According to the U.S. Food and Drug Administration (FDA), it is "amending the food additive regulations to provide for the safe use of ozone in gaseous and aqueous phases as an anti-microbial agent on food, including meat and poultry." Studies at California Polytechnic University demonstrated that 0.3 µmol/mol levels of ozone dissolved in filtered tapwater can produce a reduction of more than 99.99% in such food-borne microorganisms as salmonella, *E. coli* 0157:H7 and *Campylobacter*. This quantity is 20,000 times the WHO-recommended limits stated above. Ozone can be used to remove pesticide residues from fruits and vegetables.

Ozone is used in homes and hot tubs to kill bacteria in the water and to reduce the amount of chlorine or bromine required by reactivating them to their free state. Since ozone does not remain in the water long enough, ozone by itself is ineffective at preventing cross-contamination among bathers and must be used in conjunction with halogens. Gaseous ozone created by ultraviolet light or by corona discharge is injected into the water.

Ozone is also widely used in treatment of water in aquariums and fish ponds. Its use can minimize bacterial growth, control parasites, eliminate transmission of some diseases, and reduce or eliminate "yellowing" of the water. Ozone must not come in contact with fish's gill structures. Natural salt water (with life forms) provides enough "instantaneous demand" that controlled amounts of ozone activate bromide ion to hypobromous acid, and the ozone entirely decays in a few seconds to minutes. If oxygen fed ozone is used, the water will be higher in dissolved oxygen, fish's gill structures will atrophy and they will become dependent on higher dissolved oxygen levels.

Aquaculture

Ozonation - a process of infusing water with ozone - can be used in aquaculture to facilitate organic breakdown. Ozone is also added to recirculating systems to reduce nitrite levels through conversion into nitrate. If nitrite levels in the water are high, nitrites will also accumulate in the blood and tissues of fish, where it interferes with oxygen transport (it causes oxidation of the heme-group of haemoglobin from ferrous ($Fe2+$) to ferric ($Fe3+$), making haemoglobin unable to bind $O2$). Despite these apparent positive effects, ozone use in recirculation systems has been linked to reducing the level of bioavailable iodine in salt water systems, resulting in iodine deficiency symptoms such as goitre and decreased growth in Senegalese sole (Solea senegalensis) larvae.

Ozonate seawater is used for surface disinfection of haddock and Atlantic halibut eggs against nodavirus. Nodavirus is a lethal and vertically transmitted virus which causes severe mortality in fish. Haddock eggs should not be treated with high ozone level as eggs so treated did not hatch and died after 3–4 days.

Agriculture

Ozone application on freshly cut pineapple and banana shows increase in flavonoids and total phenol contents when exposure is up to 20 minutes. Decrease in ascorbic acid (one form of vitamin C) content is observed but the positive effect on total phenol content and flavonoids can overcome the negative effect. Tomatoes upon treatment with ozone shows an increase in β-carotene, lutein and lycopene. However, ozone application on strawberries in pre-harvest period shows decrease in ascorbic acid content.

Ozone facilitates the extraction of some heavy metals from soil using EDTA. EDTA forms strong, water-soluble coordination compounds with some heavy metals (Pb, Zn) thereby making it possible to dissolve them out from contaminated soil. If contaminated soil is pre-treated with ozone, the extraction efficacy of Pb, Am and Pu increases by 11.0–28.9%, 43.5% and 50.7% respectively.

Medical

Various therapeutic uses for Ozone have been proposed, but are not supported by peer-reviewed evidence and generally considered alternative medicine.

Chlorofluorocarbon

A chlorofluorocarbon (CFC) is an organic compound that contains only carbon, chlorine, and fluorine, produced as volatile derivative of methane, ethane, and propane. They are also commonly known by the DuPont brand name **Freon**. The most common representative is dichlorodifluoromethane (R-12 or Freon-12). Many CFCs have been widely used as refrigerants, propellants (in aerosol applications), and solvents. Because CFCs contribute to ozone depletion in the upper atmosphere, the manufacture of such compounds has been phased out under the Montreal Protocol, and they are being replaced with other products such as hydrofluorocarbons (HFCs) (e.g., R-410A) and R-134a.

Structure, Properties and Production

As in simpler alkanes, carbon in the CFCs bonds with tetrahedral symmetry. Because the fluorine and chlorine atoms differ greatly in size and effective charge from hydrogen and from each other, the methane-derived CFCs deviate from perfect tetrahedral symmetry.

The physical properties of CFCs and HCFCs are tunable by changes in the number and identity of the halogen atoms. In general they are volatile, but less so than their parent alkanes. The decreased volatility is attributed to the molecular polarity induced by the halides, which induces intermolecular interactions. Thus, methane boils at −161 °C whereas the fluoromethanes boil between −51.7 (CF_2H_2) and −128 °C (CF_4). The CFCs have still higher boiling points because the chloride is even more polarizable than fluoride. Because of their polarity, the CFCs are useful solvents, and their boiling points make them suitable as refrigerants. The CFCs are far less flammable than methane, in part because they contain fewer C-H bonds and in part because, in the case of the chlorides and bromides, the released halides quench the free radicals that sustain flames.

The densities of CFCs are higher than their corresponding alkanes. In general the density of these compounds correlates with the number of chlorides.

CFCs and HCFCs are usually produced by halogen exchange starting from chlorinated methanes and ethanes. Illustrative is the synthesis of chlorodifluoromethane from chloroform:

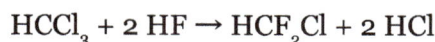

$$HCCl_3 + 2\ HF \rightarrow HCF_2Cl + 2\ HCl$$

The brominated derivatives are generated by free-radical reactions of the chlorofluorocarbons, replacing C-H bonds with C-Br bonds. The production of the anesthetic 2-bromo-2-chloro-1,1,1-trifluoroethane ("halothane") is illustrative:

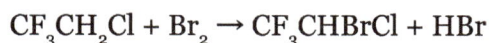

$$CF_3CH_2Cl + Br_2 \rightarrow CF_3CHBrCl + HBr$$

Reactions

The most important reaction of the CFCs is the photo-induced scission of a C-Cl bond:

$$CCl_3F \rightarrow CCl_2F\cdot + Cl\cdot$$

The chlorine atom, written often as $Cl\cdot$, behaves very differently from the chlorine molecule (Cl_2). The radical $Cl\cdot$ is long-lived in the upper atmosphere, where it catalyzes the conversion of ozone into O_2. Ozone absorbs UV-B radiation, so its depletion allows more of this high energy radiation to reach the Earth's surface. Bromine atoms are even more efficient catalysts; hence brominated CFCs are also regulated.

Applications

Applications exploit the low toxicity, low reactivity, and low flammability of the CFCs and HCFCs. Every permutation of fluorine, chlorine, and hydrogen based on methane and ethane has been examined and most have been commercialized. Furthermore, many examples are known for higher numbers of carbon as well as related compounds containing bromine. Uses include refrigerants, blowing agents, propellants in medicinal applications, and degreasing solvents.

Billions of kilograms of chlorodifluoromethane are produced annually as precursor to tetrafluoroethylene, the monomer that is converted into Teflon.

Classes of Compounds, Nomenclature

- Chlorofluorocarbons (CFCs): when derived from methane and ethane these compounds have the formulae CCl_mF_{4-m} and $C_2Cl_mF_{6-m}$, where m is nonzero.

- Hydro-chlorofluorocarbons (HCFCs): when derived from methane and ethane these compounds have the formula $CCl_mF_nH_{4-m-n}$ and $C_2Cl_xF_yH_{6-x-y}$, where m, n, x, and y are nonzero.

- and bromofluorocarbons have formulae similar to the CFCs and HCFCs but also include bromine.

- Hydrofluorocarbons (HFCs): when derived from methane, ethane, propane, and butane, these compounds have the respective formulae CF_mH_{4-m}, $C_2F_mH_{6-m}$, $C_3F_mH_{8-m}$, and $C_4F_mH_{10-m}$, where m is nonzero.

Numbering System

A numbering system is used for fluorinated alkanes, prefixed with Freon-, R-, CFC-, and HCFC-. The rightmost value indicates the number of fluorine atoms, the next value to the left is the number of hydrogen atoms *plus* 1, and the next value to the left is the number of carbon atoms *less* one (zeroes are not stated). Remaining atoms are chlorine. Thus, Freon-12 indicates a methane derivative (only two numbers) containing two fluorine atoms (the second 2) and no hydrogen (1-1=0). It is therefore CCl_2F_2.

Another, easier equation that can be applied to get the correct molecular formula of the CFC/R/Freon class compounds is this to take the numbering and add 90 to it. The resulting value will give the number of carbons as the first numeral, the second numeral gives the number of hydrogen atoms, and the third numeral gives the number of fluorine atoms. The rest of the unaccounted carbon bonds are occupied by chlorine atoms. The value of this equation is always a three figure number. An easy example is that of CFC-12, which gives: 90+12=102 -> 1 carbon, 0 hydrogens, 2 fluorine atoms, and hence 2 chlorine atoms resulting in CCl_2F_2. The main advantage of this method of deducing the molecular composition in comparison with the method described in the paragraph above is that it gives the number of carbon atoms of the molecule.

Freons containing bromine are signified by four numbers. Isomers, which are common for ethane and propane derivatives, are indicated by letters following the numbers.

Principal CFCs			
Systematic name	Common/trivial name(s), code	Boiling point (°C)	Formula
Trichlorofluoromethane	Freon-11, R-11, CFC-11	23.77	CCl_3F
Dichlorodifluoromethane	Freon-12, R-12, CFC-12	−29.8	CCl_2F_2
Difluoromethane/pentafluoroethane	R-410A, Puron, AZ-20	−48.5	50% CH_2F_2/50% CHF_2CF_3
Chlorotrifluoromethane	Freon-13, R-13, CFC-13	−81	$CClF_3$
Chlorodifluoromethane	R-22, HCFC-22	−40.8	$CHClF_2$
Dichlorofluoromethane	R-21, HCFC-21	8.9	$CHCl_2F$
Chlorofluoromethane	Freon 31, R-31, HCFC-31	−9.1	CH_2ClF
Bromochlorodifluoromethane	BCF, Halon 1211, H-1211, Freon 12B1	−3.7	$CBrClF_2$

1,1,2-Trichloro-1,2,2-trifluoroethane	Freon 113, R-113, CFC-113, 1,1,2-Trichlorotrifluoroethane	47.7	$Cl_2FC\text{-}CClF_2$
1,1,1-Trichloro-2,2,2-trifluoroethane	Freon 113a, R-113a, CFC-113a	45.9	$Cl_3C\text{-}CF_3$
1,2-Dichloro-1,1,2,2-tetrafluoroethane	Freon 114, R-114, CFC-114, Dichlorotetrafluoroethane	3.8	$ClF_2C\text{-}CClF_2$
1-Chloro-1,1,2,2,2-pentafluoroethane	Freon 115, R-115, CFC-115, Chloropentafluoroethane	−38	$ClF_2C\text{-}CF_3$
2-Chloro-1,1,1,2-tetrafluoroethane	R-124, HCFC-124	−12	$CHFClCF_3$
1,1-Dichloro-1-fluoroethane	R-141b, HCFC-141b	32	$Cl_2FC\text{-}CH_3$
1-Chloro-1,1-difluoroethane	R-142b, HCFC-142b	−9.2	$ClF_2C\text{-}CH_3$
Tetrachloro-1,2-difluoroethane	Freon 112, R-112, CFC-112	91.5	CCl_2FCCl_2F
Tetrachloro-1,1-difluoroethane	Freon 112a, R-112a, CFC-112a	91.5	$CClF_2CCl_3$
1,1,2-Trichlorotrifluoroethane	Freon 113, R-113, CFC-113	48	CCl_2FCClF_2
1-bromo-2-chloro-1,1,2-trifluoroethane	Halon 2311a	51.7	$CHClFCBrF_2$
2-bromo-2-chloro-1,1,1-trifluoroethane	Halon 2311	50.2	$CF_3CHBrCl$
1,1-Dichloro-2,2,3,3,3-pentafluoropropane	R-225ca, HCFC-225ca	51	$CF_3CF_2CHCl_2$
1,3-Dichloro-1,2,2,3,3-pentafluoropropane	R-225cb, HCFC-225cb	56	$CClF_2CF_2CHClF$

History

Carbon tetrachloride (CCl_4) was used in fire extinguishers and glass "anti-fire grenades" from the late nineteenth century until around the end of World War II. Experimentation with chloroalkanes for fire suppression on military aircraft began at least as early as the 1920s. *Freon* is a trade name for a group of CFCs which are used primarily as refrigerants, but also have uses in fire-fighting and as propellants in aerosol cans. Bromomethane is widely used as a fumigant. Dichloromethane is a versatile industrial solvent.

The Belgian scientist Frédéric Swarts pioneered the synthesis of CFCs in the 1890s. He developed an effective exchange agent to replace chloride in carbon tetrachloride with fluoride to synthesize CFC-11 (CCl_3F) and CFC-12 (CCl_2F_2).

In the late 1920s, Thomas Midgley, Jr. improved the process of synthesis and led the effort to use CFC as refrigerant to replace ammonia (NH_3), chloromethane (CH_3Cl), and sulfur dioxide (SO_2), which are toxic but were in common use. In searching for a new refrigerant, requirements for the compound were: low boiling point, low toxicity, and to be generally non-reactive. In a demonstration for the American Chemical Society, Midgley flamboyantly demonstrated all these properties by inhaling a breath of the gas and using it to blow out a candle in 1930.

Commercial Development and Use

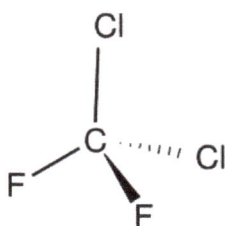

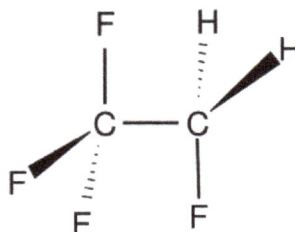

Dichlorodifluoromethane   Tetrafluorethane

During World War II, various chloroalkanes were in standard use in military aircraft, although these early halons suffered from excessive toxicity. Nevertheless, after the war they slowly became more common in civil aviation as well. In the 1960s, fluoroalkanes and bromofluoroalkanes became available and were quickly recognized as being highly effective fire-fighting materials. Much early research with Halon 1301 was conducted under the auspices of the US Armed Forces, while Halon 1211 was, initially, mainly developed in the UK. By the late 1960s they were standard in many applications where water and dry-powder extinguishers posed a threat of damage to the protected property, including computer rooms, telecommunications switches, laboratories, museums and art collections. Beginning with warships, in the 1970s, bromofluoroalkanes also progressively came to be associated with rapid knockdown of severe fires in confined spaces with minimal risk to personnel.

By the early 1980s, bromofluoroalkanes were in common use on aircraft, ships, and large vehicles as well as in computer facilities and galleries. However, concern was beginning to be expressed about the impact of chloroalkanes and bromoalkanes on the ozone layer. The Vienna Convention for the Protection of the Ozone Layer did not cover bromofluoroalkanes as it was thought, at the time, that emergency discharge of extinguishing systems was too small in volume to produce a significant impact, and too important to human safety for restriction.

Regulation

Since the late 1970s, the use of CFCs has been heavily regulated because of their destructive effects on the ozone layer. After the development of his electron capture detector, James Lovelock was the first to detect the widespread presence of CFCs in the air, finding a mole fraction of 60 ppt of CFC-11 over Ireland. In a self-funded research expedition ending in 1973, Lovelock went on to measure CFC-11 in both the Arctic and Antarctic, finding the presence of the gas in each of 50 air samples collected, and concluding that CFCs are not hazardous to the environment. The experiment did however provide the first useful data on the presence of CFCs in the atmosphere. The damage caused by CFCs was discovered by Sherry Rowland and Mario Molina who, after hearing a lecture on the subject of Lovelock's work, embarked on research resulting in the first publication suggesting the connection in 1974. It turns out that one of CFCs' most attractive features—their low reactivity— is key to their most destructive effects. CFCs' lack of reactivity gives them a lifespan that can exceed 100 years, giving them time to diffuse into the upper stratosphere. Once in the stratosphere, the sun's ultraviolet radiation is strong enough to cause the homolytic cleavage of the C-Cl bond.

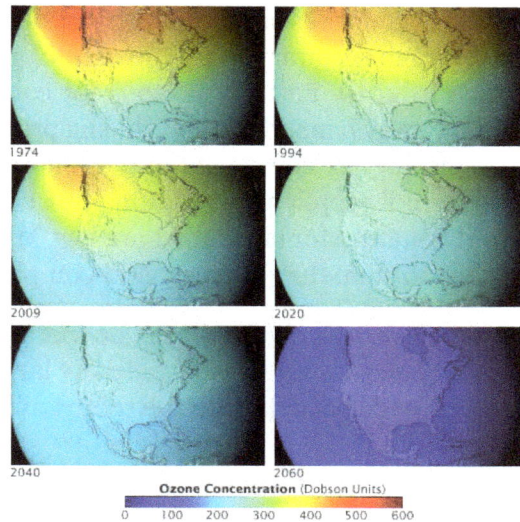

NASA projection of stratospheric ozone, in Dobson units, if chlorofluorocarbons had not been banned. Animated version.

By 1987, in response to a dramatic seasonal depletion of the ozone layer over Antarctica, diplomats in Montreal forged a treaty, the Montreal Protocol, which called for drastic reductions in the production of CFCs. On 2 March 1989, 12 European Community nations agreed to ban the production of all CFCs by the end of the century. In 1990, diplomats met in London and voted to significantly strengthen the Montreal Protocol by calling for a complete elimination of CFCs by the year 2000. By the year 2010 CFCs should have been completely eliminated from developing countries as well.

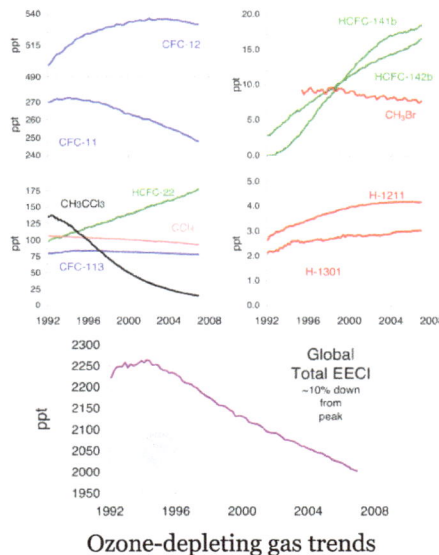

Ozone-depleting gas trends

Because the only CFCs available to countries adhering to the treaty is from recycling, their prices have increased considerably. A worldwide end to production should also terminate the smuggling of this material. However, there are current CFC smuggling issues, as recognized by the United Nations Environmental Programme (UNEP) in a 2006 report titled "Illegal Trade in Ozone Depleting Substances". UNEP estimates that between 16,000–38,000 tonnes of CFCs passed through the black market in the mid-1990s. The report estimated between 7,000 and 14,000 tonnes of CFCs are smuggled annually into developing countries. Asian countries are those with the most smuggling; as of 2007, China, India and South Korea were found to account for around 70% of global CFC production, South Korea later to ban CFC production in 2010. Possible reasons for continued CFC smuggling were also examined: the report noted that many banned CFC producing products have long lifespans and continue to operate. The cost of replacing the equipment of these items is sometimes cheaper than outfitting them with a more ozone-friendly appliance. Additionally, CFC smuggling is not considered a significant issue, so the perceived penalties for smuggling are low. While the eventual phaseout of CFCs is likely, efforts are being taken to stem these current non-compliance problems.

By the time of the Montreal Protocol it was realised that deliberate and accidental discharges during system tests and maintenance accounted for substantially larger volumes than emergency discharges, and consequently halons were brought into the treaty, albeit with many exceptions.

Regulatory Gap

While the production and consumption of CFCs are regulated under the Montreal Protocol, emissions from existing banks of CFCs are not regulated under the agreement. In 2002, there were an estimated 5,791 kilotons of CFCs in existing products such as refrigerators, air conditioners, aerosol cans and others. Approximately one-third of these CFCs are projected to be emitted over the next decade if action is not taken, posing a threat to both the ozone layer and the climate. A proportion of these CFCs can be safely captured and destroyed.

Regulation and DuPont

In 1978 the United States banned the use of CFCs such as Freon in aerosol cans, the beginning of a long series of regulatory actions against their use. The critical DuPont manufacturing patent for Freon ("Process for Fluorinating Halohydrocarbons", U.S. Patent #3258500) was set to expire in 1979. In conjunction with other industrial peers DuPont sponsored efforts such as the "Alliance for Responsible CFC Policy" to question anti-CFC science, but in a turnabout in 1986 DuPont, with new patents in hand, publicly condemned CFCs. DuPont representatives appeared before the Montreal Protocol urging that CFCs be banned worldwide and stated that their new HCFCs would meet the worldwide demand for refrigerants.

Phasing-out of CFCs

Use of certain chloroalkanes as solvents for large scale application, such as dry cleaning, have been phased out, for example, by the IPPC directive on greenhouse gases in 1994 and by the volatile organic compounds (VOC) directive of the EU in 1997. Permitted chlorofluoroalkane uses are medicinal only.

Bromofluoroalkanes have been largely phased out and the possession of equipment for their use is prohibited in some countries like the Netherlands and Belgium, from 1 January 2004, based on the Montreal Protocol and guidelines of the European Union.

Production of new stocks ceased in most (probably all) countries in 1994. However many countries still require aircraft to be fitted with halon fire suppression systems because no safe and completely satisfactory alternative has been discovered for this application. There are also a few other, highly specialized uses. These programs recycle halon through "halon banks" coordinated by the Halon Recycling Corporation to ensure that discharge to the atmosphere occurs only in a genuine emergency and to conserve remaining stocks.

The interim replacements for CFCs are hydrochlorofluorocarbons (HCFCs), which deplete stratospheric ozone, but to a much lesser extent than CFCs. Ultimately, hydrofluorocarbons (HFCs) will replace HCFCs. Unlike CFCs and HCFCs, HFCs have an ozone depletion potential (ODP) of 0. DuPont began producing hydrofluorocarbons as alternatives to Freon in the 1980s. These included Suva refrigerants and Dymel propellants. Natural refrigerants are climate friendly solutions that are enjoying increasing support from large companies and governments interested in reducing global warming emissions from refrigeration and air conditioning. Hydrofluorocarbons are included in the Kyoto Protocol because of their very high Global Warming Potential and are facing calls to be regulated under the Montreal Protocol due to the recognition of halocarbon contributions to climate change.

On 21 September 2007, approximately 200 countries agreed to accelerate the elimination of hydrochlorofluorocarbons entirely by 2020 in a United Nations-sponsored Montreal summit. Developing nations were given until 2030. Many nations, such as the United States and China, who had previously resisted such efforts, agreed with the accelerated phase out schedule.

Development of Alternatives for CFCs

Work on alternatives for chlorofluorocarbons in refrigerants began in the late 1970s after the first warnings of damage to stratospheric ozone were published.

The hydrochlorofluorocarbons (HCFCs) are less stable in the lower atmosphere, enabling them to break down before reaching the ozone layer. Nevertheless, a significant fraction of the HCFCs do break down in the stratosphere and they have contributed to more chlorine buildup there than originally predicted. Later alternatives lacking the chlorine, the hydrofluorocarbons (HFCs) have an even shorter lifetimes in the lower atmosphere. One of these compounds, HFC-134a, is now used in place of CFC-12 in automobile air conditioners. Hydrocarbon refrigerants (a propane/isobutane blend) are also used extensively in mobile air conditioning systems in Australia, the USA and many other countries, as they have excellent thermodynamic properties and perform particularly well in high ambient temperatures.

One of the natural refrigerants (along with ammonia and carbon dioxide), hydrocarbons have negligible environmental impacts and are also used worldwide in domestic and commercial refrigeration applications, and are becoming available in new split system air conditioners. Various other solvents and methods have replaced the use of CFCs in laboratory analytics.

Applications and replacements for CFCs		
Application	**Previously used CFC**	**Replacement**
Refrigeration & air-conditioning	CFC-12 (CCl_2F_2); CFC-11(CCl_3F); CFC-13($CClF_3$); HCFC-22 ($CHClF_2$); CFC-113 ($Cl_2FCCClF_2$); CFC-114 ($CClF_2CClF_2$); CFC-115 (CF_3CClF_2);	HFC-23 (CHF_3); HFC-134a (CF_3CFH_2); HFC-507 (a 1:1 azeotropic mixture of HFC 125 (CF_3CHF2) and HFC-143a (CF_3CH_3)); HFC 410 (a 1:1 azeotropic mixture of HFC-32 (CF_2H_2) and HFC-125 (CF_3CF_2H))

Propellants in medicinal aerosols	CFC-114 ($CClF_2CClF_2$)	HFC-134a (CF_3CFH_2); HFC-227ea (CF_3CHFCF_3)
Blowing agents for foams	CFC-11 (CCl_3F); CFC 113 ($Cl_2FCCClF_2$); HCFC-141b (CCl_2FCH_3)	HFC-245fa ($CF_3CH_2CHF_2$); HFC-365 mfc ($CF_3CH_2CF_2CH_3$)
Solvents, degreasing agents, cleaning agents	CFC-11 (CCl_3F); CFC-113 (CCl_2FCClF_2)	None

Environmental Impacts

As previously discussed, CFCs were phased out via the Montreal Protocol due to their part in ozone depletion. However, the atmospheric impacts of CFCs are not limited to its role as an active ozone reducer. This anthropogenic compound is also a greenhouse gas, with a much higher potential to enhance the greenhouse effect than CO_2.

Infrared absorption bands trap heat from escaping earth's atmosphere. In the case of CFCs, the strongest of these bands are located in the spectral region 7.8–15.3 μm – referred to as an atmospheric window due to the relative transparency of the atmosphere within this region. The strength of CFC bands and the unique susceptibility of the atmosphere, at which the compound absorbs and emits radiation, are two factors that contribute to CFCs' "super" greenhouse effect. Another such factor is the low concentration of the compound. Because CO_2 is close to saturation with high concentrations, it takes more of the substance to enhance the greenhouse effect. Conversely, the low concentration of CFCs allow their effects to increase linearly with mass.

Tracer of Ocean Circulation

Because the time history of CFC concentrations in the atmosphere is relatively well known, they have provided an important constraint on ocean circulation. CFCs dissolve in seawater at the ocean surface and are subsequently transported into the ocean interior. Because CFCs are inert, their concentration in the ocean interior reflects simply the convolution of their atmospheric time evolution and ocean circulation and mixing.

CFC and SF_6 Tracer-Derived Age of Ocean Water

Chlorofluorocarbons (CFCs) are anthropogenic compounds that have been released into the atmosphere since the 1930s in various applications such as in air-conditioning, refrigeration, blowing agents in foams, insulations and packing materials, propellants in aerosol cans, and as solvents. The entry of CFCs into the ocean makes them extremely useful as transient tracers to estimate rates and pathways of ocean circulation and mixing processes. However, due to production restrictions of CFCs in the 1980s, atmospheric concentrations of CFC-11 and CFC-12 has stopped increasing, and the CFC-11 to CFC-12 ratio in the atmosphere have been steadily decreasing, making water dating of water masses more problematic. Incidentally, production and release of sulfur hexafluoride (SF_6) have rapidly increased in the atmosphere since the 1970s. Similar to CFCs, SF_6 is also an inert gas and is not affected by oceanic chemical or biological activities. Thus, using CFCs in concert with SF_6 as a tracer resolves the water dating issues due to decreased CFC concentrations.

Using CFCs or SF_6 as a tracer of ocean circulation allows for the derivation of rates for ocean processes due to the time-dependent source function. The elapsed time since a subsurface water mass was last in contact with the atmosphere is the tracer-derived age. Estimates of age can be derived based on the partial pressure of an individual compound and the ratio of the partial pressure of CFCs to each other (or SF_6).

Partial Pressure and Ratio Dating Techniques

The age of a water parcel can be estimated by the CFC partial pressure (pCFC) age or SF_6 partial pressure (pSF_6) age. The pCFC age of a water sample is defined as:

pCFC = [CFC]/F(T,S)

where [CFC] is the measured CFC concentration (pmol kg^{-1}) and F is the solubility of CFC gas in seawater as a function of temperature and salinity. The CFC partial pressure is expressed in units of 10−12 atmospheres or parts-per-trillion (ppt). The solubility measurements of CFC-11 and CFC-12 have been previously measured by Warner and Weiss Additionally, the solubility measurement of CFC-113 was measured by Bu and Warner and SF_6 by Wanninkhof et al. and Bullister et al. Theses authors mentioned above have expressed the solubility, F, at a total pressure of 1 atm as:

$$\ln F = a_1 + a_2(100/T) + a_3\ln(T/100) + a_4(T/100)^2 + S[b_1 + b_2(T/100) + b_3(T/100)^2]$$

where F = solubility expressed in either mol l^{-1} or mol kg^{-1} atm^{-1} T = absolute temperature S = salinity in parts per thousand (ppt) a_1, a_2, a_3, b_1, b_2, and b_3 are constants to be determined from the least squares fit to the solubility measurements. This equation is derived from the integrated Van 't Hoff equation and the logarithmic Setchenow salinity dependence.

It can be noted that the solubility of CFCs increase with decreasing temperature at approximately 1% per degree Celsius.

Once the partial pressure of the CFC (or SF_6) is derived, it is then compared to the atmospheric time histories for CFC-11, CFC-12, or SF_6 in which the pCFC directly corresponds to the year with the same. The difference between the corresponding date and the collection date of the seawater sample is the average age for the water parcel. The age of a parcel of water can also be calculated using the ratio of two CFC partial pressures or the ratio of the SF_6 partial pressure to a CFC partial pressure.

Safety

According to their material safety data sheets, CFCs and HCFCs are colorless, volatile, toxic liquids and gases with a faintly sweet ethereal odor. Overexposure at concentrations of 11% or more may cause dizziness, loss of concentration, central nervous system depression and/or cardiac arrhythmia. Vapors displace air and can cause asphyxiation in confined spaces. Although non-flammable, their combustion products include hydrofluoric acid, and related species. Normal occupational exposure is rated at 0.07% and does not pose any serious health risks.

References

- Wallace, John M. and Peter V. Hobbs. Atmospheric Science; An Introductory Survey.Elsevier. Second Edition, 2006. ISBN 978-0-12-732951-2.

- Evans, Kimberly Masters (2005). "The greenhouse effect and climate change". The environment: a revolution in attitudes. Detroit: Thomson Gale. ISBN 0-7876-9082-1.

- Bridging the Emissions Gap: A UNEP Synthesis Report (PDF), Nairobi, Kenya: United Nations Environment Programme (UNEP), November 2011, ISBN 978-92-807-3229-0.

- Donald G. Kaufman; Cecilia M. Franz (1996). Biosphere 2000: protecting our global environment. Kendall/Hunt Pub. Co. ISBN 978-0-7872-0460-0. Retrieved 11 October 2011.

- Ocean Acidification: A National Strategy to Meet the Challenges of a Changing Ocean. Washington, DC: National Academies Press. doi:10.17226/12904. ISBN 978-0-309-15359-1.

- Strassburger, Julius (1969). Blast Furnace Theory and Practice. New York: American Institute of Mining, Metallurgical, and Petroleum Engineers. ISBN 0-677-10420-0.

- Drysdale, Dougal (2008). "Physics and Chemistry of Fire". In Cote, Arthur E. Fire Protection Handbook. 1 (20th ed.). Quincy, MA: National Fire Protection Association. pp. 2–18. ISBN 978-0-87765-758-3.

- Miller, Ron; Hartmann, William K. (2005). The Grand Tour: A Traveler's Guide to the Solar System (3rd ed.). Thailand: Workman Publishing. pp. 172–73. ISBN 0-7611-3547-2.

- Sneader W (2005). Drug Discovery –A History. (Part 1: Legacy of the past, chapter 8: systematic medicine, pp. 74–87). John Wiley and Sons. ISBN 978-0-471-89980-8. Retrieved 21 April 2010.

- Housecroft, Catherine E. & Sharpe, Alan G. (2008). "Chapter 15: The group 15 elements". Inorganic Chemistry (3rd ed.). Pearson. p. 464. ISBN 978-0-13-175553-6.

- Brown, Theodore L.; LeMay, H. Eugene, Jr.; Bursten, Bruce E.; Burdge, Julia R. (2003) [1977]. "22". In Nicole Folchetti. Chemistry: The Central Science (9th ed.). Pearson Education. pp. 882–883. ISBN 0-13-066997-0.

Consequences of Global Warming

Climate change can cause significant damage to humans, the environment and animals. All living things depend on a fine balance of natural interventions, such as high levels of rainfall or drought as well as predictable, seasonal and hospitable climates. Global warming can alter these much-desired phenomena for the worse. Global warming is best understood in confluence with the major topics listed in the following chapter.

Effects of Global Warming

The effects of global warming are the environmental and social changes caused (directly or indirectly) by human emissions of greenhouse gases. There is a scientific consensus that climate change is occurring, and that human activities are the primary driver. Many impacts of climate change have already been observed, including glacier retreat, changes in the timing of seasonal events (e.g., earlier flowering of plants), and changes in agricultural productivity.

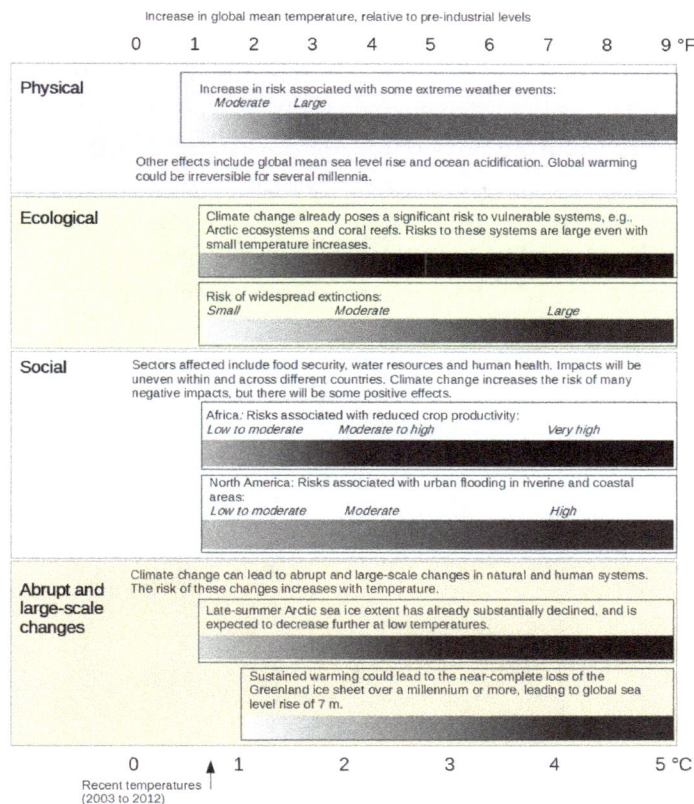

Increase in global mean temperature, relative to pre-industrial levels

| | 0 | 1 | 2 | 3 | 4 | 5 | 6 | 7 | 8 | 9 °F |

Physical
Increase in risk associated with some extreme weather events:
Moderate   *Large*

Other effects include global mean sea level rise and ocean acidification. Global warming could be irreversible for several millennia.

Ecological
Climate change already poses a significant risk to vulnerable systems, e.g., Arctic ecosystems and coral reefs. Risks to these systems are large even with small temperature increases.

Risk of widespread extinctions:
Small   *Moderate*   *Large*

Social
Sectors affected include food security, water resources and human health. Impacts will be uneven within and across different countries. Climate change increases the risk of many negative impacts, but there will be some positive effects.

Africa: Risks associated with reduced crop productivity:
Low to moderate   *Moderate to high*   *Very high*

North America: Risks associated with urban flooding in riverine and coastal areas:
Low to moderate   *Moderate*   *High*

Abrupt and large-scale changes
Climate change can lead to abrupt and large-scale changes in natural and human systems. The risk of these changes increases with temperature.

Late-summer Arctic sea ice extent has already substantially declined, and is expected to decrease further at low temperatures.

Sustained warming could lead to the near-complete loss of the Greenland ice sheet over a millennium or more, leading to global sea level rise of 7 m.

| | 0 | ↑ 1 | 2 | 3 | 4 | 5 °C |

Recent temperatures (2003 to 2012)

Summary of climate change impacts.

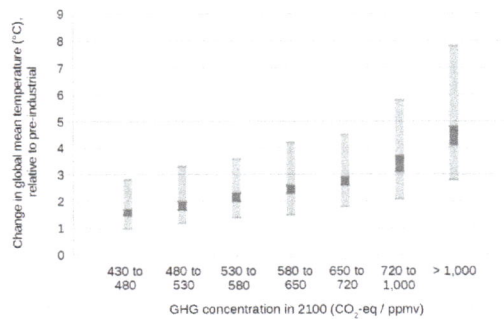

Projected global warming in 2100 for a range of emission scenarios.

Future effects of climate change will vary depending on climate change policies and social development. The two main policies to address climate change are reducing human greenhouse gas emissions (climate change mitigation) and adapting to the impacts of climate change. Geoengineering is another policy option.

Near-term climate change policies could significantly affect long-term climate change impacts. Stringent mitigation policies might be able to limit global warming (in 2100) to around 2 °C or below, relative to pre-industrial levels. Without mitigation, increased energy demand and extensive use of fossil fuels might lead to global warming of around 4 °C. Higher magnitudes of global warming would be more difficult to adapt to, and would increase the risk of negative impacts.

Definitions

In this article, "climate change" means a change in climate that persists over a sustained period of time. The World Meteorological Organization defines this time period as 30 years. Examples of climate change include increases in global surface temperature (global warming), changes in rainfall patterns, and changes in the frequency of extreme weather events. Changes in climate may be due to natural causes, e.g., changes in the sun's output, or due to human activities, e.g., changing the composition of the atmosphere. Any human-induced changes in climate will occur against a background of natural climatic variations and of variations in human activity such as population growth on shores or in arid areas which increase or decrease climate vulnerability.

Also, the term "anthropogenic forcing" refers to the influence exerted on a habitat or chemical environment by humans, as opposed to a natural process.

Temperature Changes

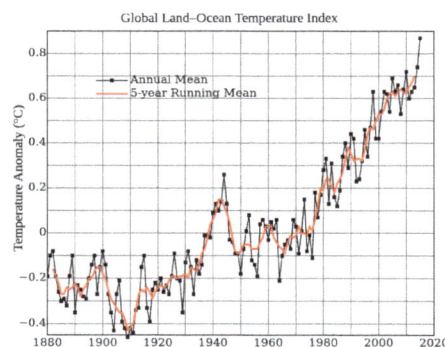

Global mean surface temperature change since 1880, relative to the 1951–1980 mean. Source: NASA GISS

The graph above shows the average of a set of temperature simulations for the 20th century (black line), followed by projected temperatures for the 21st century based on three greenhouse gas emissions scenarios (colored lines).

This article breaks down some of the impacts of climate change according to different levels of future global warming. This way of describing impacts has, for instance, been used in the IPCC (Intergovernmental Panel on Climate Change) Assessment Reports on climate change. The instrumental temperature record shows global warming of around 0.6 °C during the 20th century.

SRES Emissions Scenarios

The future level of global warming is uncertain, but a wide range of estimates (projections) have been made. The IPCC's "SRES" scenarios have been frequently used to make projections of future climate change. The SRES scenarios are "baseline" (or "reference") scenarios, which means that they do not take into account any current or future measures to limit GHG emissions (e.g., the UNFCCC's Kyoto Protocol and the Cancún agreements). Emissions projections of the SRES scenarios are broadly comparable in range to the baseline emissions scenarios that have been de-veloped by the scientific community.

In the IPCC Fourth Assessment Report, changes in future global mean temperature were projected using the six SRES "marker" emissions scenarios. Emissions projections for the six SRES "marker" scenarios are representative of the full set of forty SRES scenarios. For the lowest emissions SRES marker scenario ("B1" - see the SRES article for details on this scenario), the best estimate for global mean temperature is an increase of 1.8 °C (3.2 °F) by the end of the 21st century. This projection is relative to global temperatures at the end of the 20th century. The "likely" range (greater than 66% probability, based on expert judgement) for the SRES B1 marker scenario is 1.1–2.9 °C (2.0–5.2 °F). For the highest emissions SRES marker scenario (A1FI), the best estimate for global mean temperature increase is 4.0 °C (7.2 °F), with a "likely" range of 2.4–6.4 °C (4.3–11.5 °F).

The range in temperature projections partly reflects (1) the choice of emissions scenario, and (2) the "climate sensitivity". For (1), different scenarios make different assumptions of future social and economic development (e.g., economic growth, population level, energy policies), which in turn affects projections of greenhouse gas (GHG) emissions. The projected magnitude of warming by 2100 is closely related to the level of cumulative emissions over the 21st century (i.e. total emissions between 2000-2100). The higher the cumulative emissions over this time period, the greater the level of warming is projected to occur.

(2) reflects uncertainty in the response of the climate system to past and future GHG emissions, which is measured by the climate sensitivity). Higher estimates of climate sensitivity lead to greater projected warming, while lower estimates of climate sensitivity lead to less projected warming.

Over the next several millennia, projections suggest that global warming could be irreversible. Even if emissions were drastically reduced, global temperatures would remain close to their highest level for at least 1,000 years.

Projected Warming in Context

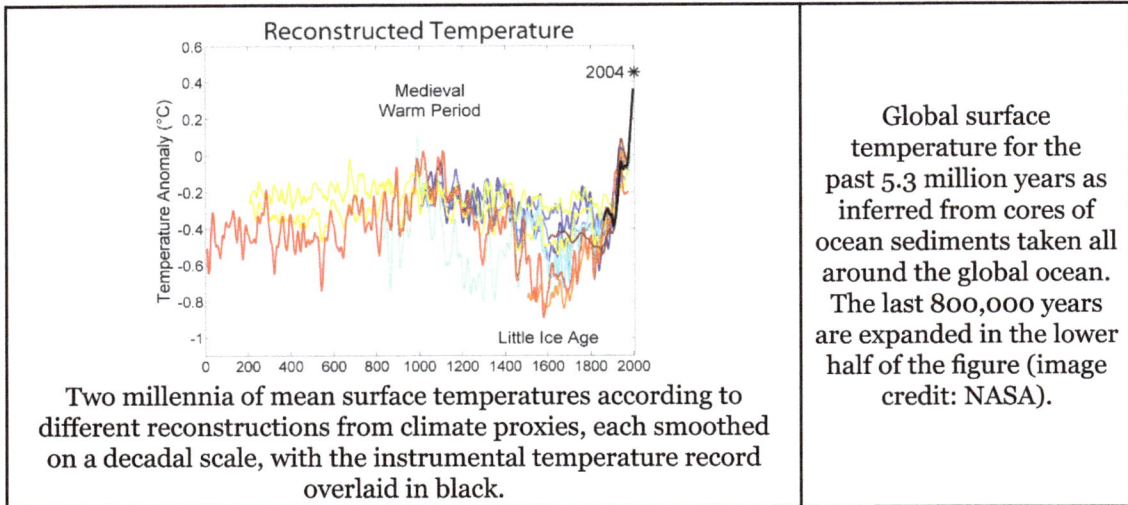

Reconstructed Temperature

Two millennia of mean surface temperatures according to different reconstructions from climate proxies, each smoothed on a decadal scale, with the instrumental temperature record overlaid in black.

Global surface temperature for the past 5.3 million years as inferred from cores of ocean sediments taken all around the global ocean. The last 800,000 years are expanded in the lower half of the figure (image credit: NASA).

Scientists have used various "proxy" data to assess past changes in Earth's climate (paleoclimate). Sources of proxy data include historical records (such as farmers' logs), tree rings, corals, fossil pollen, ice cores, and ocean and lake sediments. Analysis of these data suggest that recent warming is unusual in the past 400 years, possibly longer. By the end of the 21st century, temperatures may increase to a level not experienced since the mid-Pliocene, around 3 million years ago. At that time, models suggest that mean global temperatures were about 2–3 °C warmer than pre-industrial temperatures. Even a 2 °C rise above the pre-industrial level would be outside the range of temperatures experienced by human civilization.

Physical Impacts

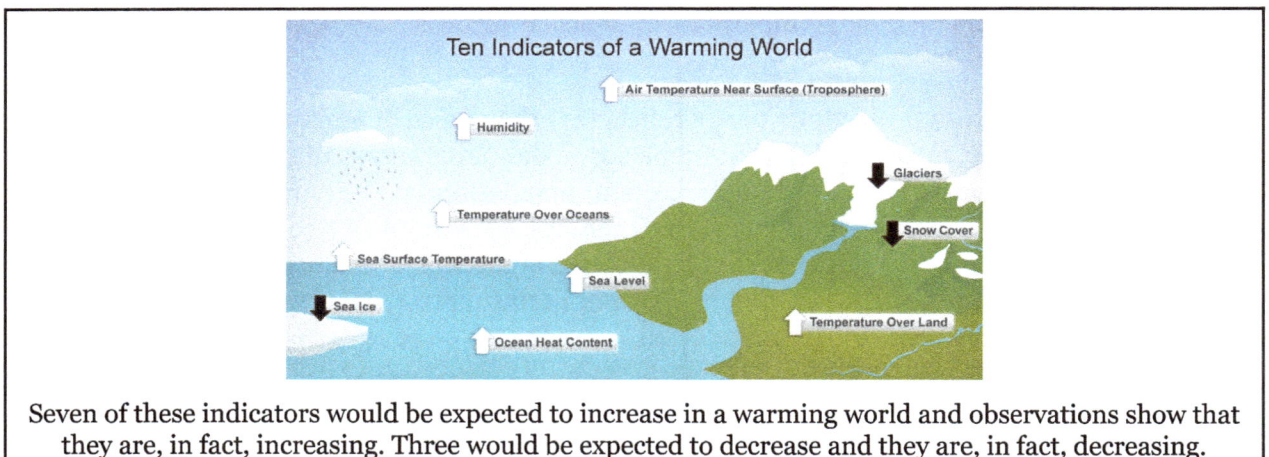

Ten Indicators of a Warming World

Seven of these indicators would be expected to increase in a warming world and observations show that they are, in fact, increasing. Three would be expected to decrease and they are, in fact, decreasing.

A broad range of evidence shows that the climate system has warmed. Evidence of global warming is shown in the graphs opposite. Some of the graphs show a positive trend, e.g., increasing temperature over land and the ocean, and sea level rise. Other graphs show a negative trend, e.g., decreased snow cover in the Northern Hemisphere, and declining Arctic sea ice extent. Evidence of warming is also apparent in living (biological) systems.

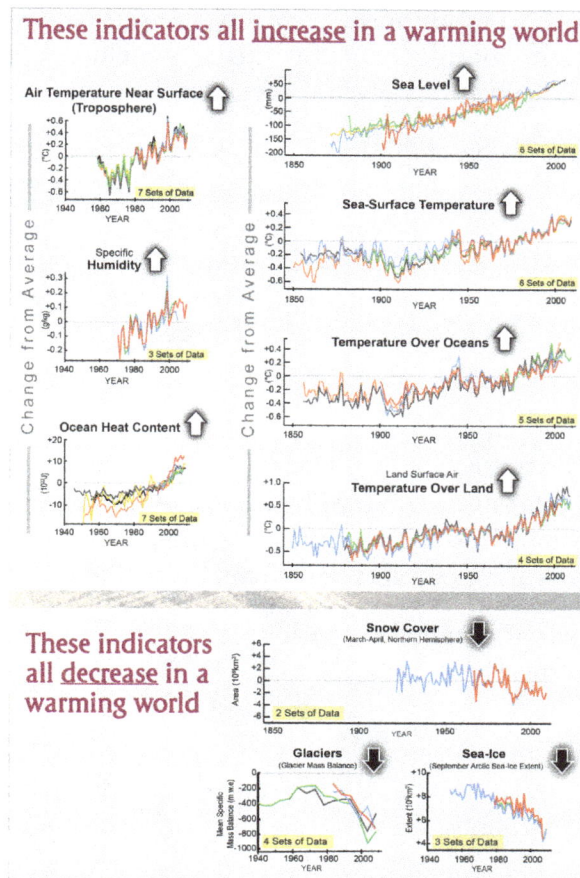

This set of graphs show changes in climate indicators over several decades. Each of the different colored lines in each panel represents an independently analyzed set of data. The data come from many different technologies including weather stations, satellites, weather balloons, ships and buoys.

Human activities have contributed to a number of the observed changes in climate. This contribution has principally been through the burning of fossil fuels, which has led to an increase in the concentration of GHGs in the atmosphere. Another human influence on the climate are sulfur dioxide emissions, which are a precursor to the formation of sulfate aerosols in the atmosphere.

Human-induced warming could lead to large-scale, irreversible, and/or abrupt changes in physical systems. An example of this is the melting of ice sheets, which contributes to sea level rise. The probability of warming having unforeseen consequences increases with the rate, magnitude, and duration of climate change.

Effects on Weather

Observations show that there have been changes in weather. As climate changes, the probabilities of certain types of weather events are affected.

Projected change in annual average precipitation by the end of the 21st century, based on a medium emissions scenario (SRES A1B) (Credit: NOAA Geophysical Fluid Dynamics Laboratory).

Changes have been observed in the amount, intensity, frequency, and type of precipitation. Widespread increases in heavy precipitation have occurred, even in places where total rain amounts have decreased. With medium confidence, IPCC (2012) concluded that human influences had contributed to an increase in heavy precipitation events at the global scale.

Projections of future changes in precipitation show overall increases in the global average, but with substantial shifts in where and how precipitation falls. Projections suggest a reduction in rainfall in the subtropics, and an increase in precipitation in subpolar latitudes and some equatorial regions. In other words, regions which are dry at present will in general become even drier, while regions that are currently wet will in general become even wetter. This projection does not apply to every locale, and in some cases can be modified by local conditions.

Extreme Weather

Over most land areas since the 1950s, it is very likely that there have been fewer or warmer cold days and nights. Hot days and nights have also very likely become warmer or more frequent. Human activities have very likely contributed to these trends. There may have been changes in other climate extremes (e.g., floods, droughts and tropical cyclones) but these changes are more difficult to identify.

Projections suggest changes in the frequency and intensity of some extreme weather events. Confidence in projections varies over time.

Near-term Projections (2016–2035)

Some changes (e.g., more frequent hot days) will probably be evident in the near term, while other near-term changes (e.g., more intense droughts and tropical cyclones) are more uncertain.

Long-term Projections (2081–2100)

Future climate change will be associated with more very hot days and fewer very cold days. The frequency, length and intensity of heat waves will very likely increase over most land areas. Higher growth in anthropogenic GHG emissions will be associated with larger increases in the frequency and severity of temperature extremes.

Assuming high growth in GHG emissions (IPCC scenario RCP8.5), presently dry regions may be affected by an increase in the risk of drought and reductions in soil moisture. Over most of the mid-latitude land masses and wet tropical regions, extreme precipitation events will very likely become more intense and frequent.

Tropical Cyclones

At the global scale, the frequency of tropical cyclones will probably decrease or be unchanged. Global mean tropical cyclone maximum wind speed and precipitation rates will likely increase. Changes in tropical cyclones will probably vary by region, but these variations are uncertain.

Effects of climate extremes

The impacts of extreme events on the environment and human society will vary. Some impacts will be beneficial—e.g., fewer cold extremes will probably lead to fewer cold deaths. Overall, however, impacts will probably be mostly negative.

Cryosphere

Mountain Glacier Changes Since 1970

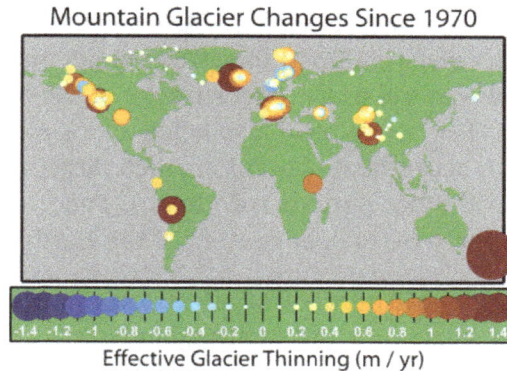

Effective Glacier Thinning (m / yr)

A map of the change in thickness of mountain glaciers since 1970. Thinning in orange and red, thickening in blue.

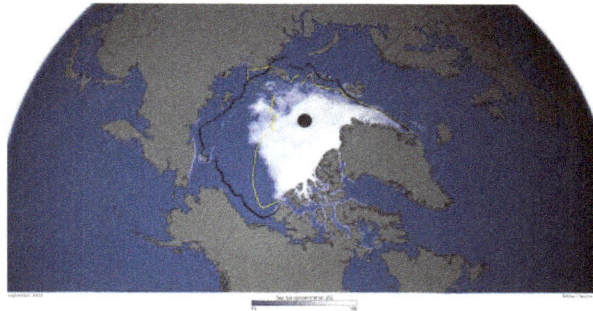

A map that shows ice concentration on 16 September 2012, along with the extent of the previous record low (yellow line) and the mid-September median extent (black line) setting a new record low that was 18 percent smaller than the previous record and nearly 50 percent smaller than the long-term (1979-2000) average.

The cryosphere is made up of areas of the Earth which are covered by snow or ice. Observed changes in the cryosphere include declines in Arctic sea ice extent, the widespread retreat of alpine glaciers, and reduced snow cover in the Northern Hemisphere.

Solomon *et al.* (2007) assessed the potential impacts of climate change on summertime Arctic sea ice extent. Assuming high growth in greenhouse gas emissions (SRES A2), some models projected that Arctic sea ice in the summer could largely disappear by the end of the 21st century. More recent projections suggest that the Arctic summers could be ice-free (defined as ice extent less than 1 million square km) as early as 2025-2030.

During the 21st century, glaciers and snow cover are projected to continue their widespread retreat. In the western mountains of North America, increasing temperatures and changes in precipitation are projected to lead to reduced snowpack. Snowpack is the seasonal accumulation of slow-melting snow. The melting of the Greenland and West Antarctic ice sheets could contribute to sea level rise, especially over long time-scales.

Changes in the cryosphere are projected to have social impacts. For example, in some regions, glacier retreat could increase the risk of reductions in seasonal water availability. Barnett *et al.* (2005) estimated that more than one-sixth of the world's population rely on glaciers and snowpack for their water supply.

Oceans

The role of the oceans in global warming is complex. The oceans serve as a sink for carbon dioxide, taking up much that would otherwise remain in the atmosphere, but increased levels of CO2 have led to ocean acidification. Furthermore, as the temperature of the oceans increases, they become less able to absorb excess CO2. The ocean have also acted as a sink in absorbing extra heat from the atmosphere. The increase in ocean heat content is much larger than any other store of energy in the Earth's heat balance over the two periods 1961 to 2003 and 1993 to 2003, and accounts for more than 90% of the possible increase in heat content of the Earth system during these periods.

Global warming is projected to have a number of effects on the oceans. Ongoing effects include rising sea levels due to thermal expansion and melting of glaciers and ice sheets, and warming of the ocean surface, leading to increased temperature stratification. Other possible effects include large-scale changes in ocean circulation.

Acidification

Changes in Aragonite Saturation of the World's Oceans, 1880–2012

Change in aragonite saturation at the ocean surface (Ω_{ar}):

| -0.8 | -0.7 | -0.6 | -0.5 | -0.4 | -0.3 | -0.2 | -0.1 | 0 |

Data source: Feely, R.A., S.C. Doney, and S.R. Cooley. 2009. Ocean acidification: Present conditions and future changes in a high-CO₂ world. Oceanography 22(4):36–47.

For more information, visit U.S. EPA's "Climate Change Indicators in the United States" at www.epa.gov/climatechange/indicators.

This map shows changes in the amount of aragonite dissolved in ocean surface waters between the 1880s and the most recent decade (2003-2012). Historical modeling suggests that since the 1880s, increased CO_2 has led to lower aragonite saturation levels (less availability of minerals) in the oceans around the world. The largest decreases in aragonite saturation have occurred in tropical waters. However, decreases in cold areas may be of greater concern because colder waters typically have lower aragonite levels to begin with.

About one-third of the carbon dioxide emitted by human activity has already been taken up by the oceans. As carbon dioxide dissolves in sea water, carbonic acid is formed, which has the effect of acidifying the ocean, measured as a change in pH. The uptake of human carbon emissions since the year 1750 has led to an average decrease in pH of 0.1 units. Projections using the SRES emissions scenarios suggest a further reduction in average global surface ocean pH of between 0.14 and 0.35 units over the 21st century.

The effects of ocean acidification on the marine biosphere have yet to be documented. Laboratory experiments suggest beneficial effects for a few species, with potentially highly detrimental effects for a substantial number of species. With medium confidence, Fischlin *et al.* (2007) projected that future ocean acidification and climate change would impair a wide range of planktonic and shallow benthic marine organisms that use aragonite to make their shells or skeletons, such as corals and marine snails (pteropods), with significant impacts particularly in the Southern Ocean.

Oxygen Depletion

The amount of oxygen dissolved in the oceans may decline, with adverse consequences for ocean life.

Sea Level Rise

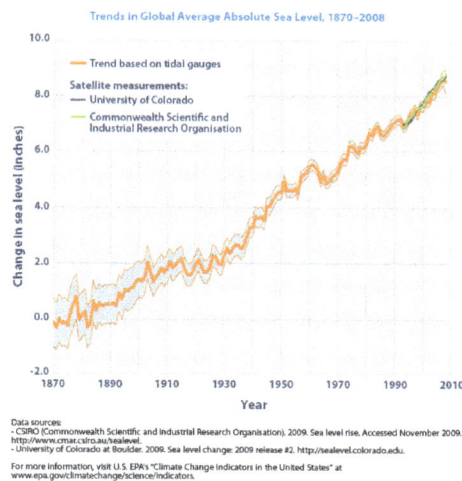

Trends in global average absolute sea level, 1870-2008.

There is strong evidence that global sea level rose gradually over the 20th century. With high confidence, Bindoff *et al.* (2007) concluded that between the mid-19th and mid-20th centuries, the rate of sea level rise increased. Authors of the IPCC Fourth Assessment Synthesis Report (IPCC AR4 SYR, 2007) reported that between the years 1961 and 2003, global average sea level rose at an average rate of 1.8 mm per year (mm/yr), with a range of 1.3–2.3 mm/yr. Between 1993 and 2003, the rate increased above the previous period to 3.1 mm/yr (range of 2.4–3.8 mm/yr). Authors of IPCC AR4 SYR (2007) were uncertain whether the increase in rate from 1993 to 2003 was due to natural variations in sea level over the time period, or whether it reflected an increase in the underlying long-term trend.

There are two main factors that have contributed to observed sea level rise. The first is thermal expansion: as ocean water warms, it expands. The second is from the contribution of land-based ice due to increased melting. The major store of water on land is found in glaciers and ice sheets. Anthropogenic forces very likely (greater than 90% probability, based on expert judgement) contributed to sea level rise during the latter half of the 20th century.

There is a widespread consensus that substantial long-term sea level rise will continue for centuries to come. In their Fourth Assessment Report, the IPCC projected sea level rise to the end of the 21st century using the SRES emissions scenarios. Across the six SRES marker scenarios, sea level was projected to rise by 18 to 59 cm (7.1 to 23.2 in), relative to sea level at the end of the 20th century. Thermal expansion is the largest component in these projections, contributing 70-75% of the central estimate for all scenarios. Due to a lack of scientific understanding, this sea level rise estimate does not include all of the possible contributions of ice sheets.

An assessment of the scientific literature on climate change was published in 2010 by the US National Research Council (US NRC, 2010). NRC (2010) described the projections in AR4 (i.e. those cited in the above paragraph) as "conservative", and summarized the results of more recent studies. Cited studies suggested a great deal of uncertainty in projections. A range of projections suggested possible sea level rise by the end of the 21st century of between 0.56 and 2 m, relative to sea levels at the end of the 20th century.

Ocean Temperature Rise

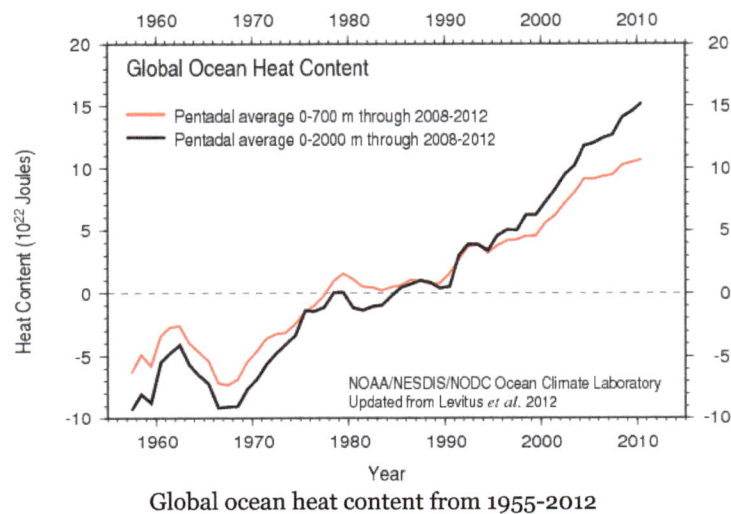

Global ocean heat content from 1955-2012

From 1961 to 2003, the global ocean temperature has risen by 0.10 °C from the surface to a depth of 700 m. There is variability both year-to-year and over longer time scales, with global ocean heat content observations showing high rates of warming for 1991–2003, but some cooling from 2003 to 2007. The temperature of the Antarctic Southern Ocean rose by 0.17 °C (0.31 °F) between the 1950s and the 1980s, nearly twice the rate for the world's oceans as a whole. As well as having effects on ecosystems (e.g. by melting sea ice, affecting algae that grow on its underside), warming reduces the ocean's ability to absorb CO_2. It is likely (greater than 66% probability, based on expert judgement) that anthropogenic forcing contributed to the general warming observed in the upper several hundred metres of the ocean during the latter half of the 20th century.

Regions

Temperatures across the world in the 1880s (left) and the 1980s (right), as compared to average temperatures from 1951 to 1980.

Projected changes in average temperatures across the world in the 2050s under three greenhouse gas (GHG) emissions scenarios.

Regional effects of global warming vary in nature. Some are the result of a generalised global change, such as rising temperature, resulting in local effects, such as melting ice. In other cases, a change may be related to a change in a particular ocean current or weather system. In such cases, the regional effect may be disproportionate and will not necessarily follow the global trend.

There are three major ways in which global warming will make changes to regional climate: melting or forming ice, changing the hydrological cycle (of evaporation and precipitation) and changing currents in the oceans and air flows in the atmosphere. The coast can also be considered a region, and will suffer severe impacts from sea level rise.

Observed Impacts

With very high confidence, Rosenzweig *et al.* (2007) concluded that physical and biological systems on all continents and in most oceans had been affected by recent climate changes, particularly regional temperature increases. Impacts include earlier leafing of trees and plants over many regions; movements of species to higher latitudes and altitudes in the Northern Hemisphere; changes in bird migrations in Europe, North America and Australia; and shifting of the oceans' plankton and fish from cold- to warm-adapted communities.

The human influence on the climate can be seen in the geographical pattern of observed warming, with greater temperature increases over land and in polar regions rather than over the oceans. Using models, it is possible to identify the human "signal" of global warming over both land and ocean areas.

Projected Impacts

Projections of future climate changes at the regional scale do not hold as high a level of scientific confidence as projections made at the global scale. It is, however, expected that future warming will follow a similar geographical pattern to that seen already, with greatest warming over land and high northern latitudes, and least over the Southern Ocean and parts of the North Atlantic Ocean. Nearly all land areas will very likely warm more than the global average.

The Arctic, Africa, small islands and Asian megadeltas are regions that are likely to be especially affected by climate change. Low-latitude, less-developed areas are at most risk of experiencing negative impacts due to climate change. Developed countries are also vulnerable to climate change. For example, developed countries will be negatively affected by increases in the severity and frequency of some extreme weather events, such as heat waves. In all regions, some people can be particularly at risk from climate change, such as the poor, young children and the elderly.

Social Systems

The impacts of climate change can be thought of in terms of sensitivity and vulnerability. "Sensitivity" is the degree to which a particular system or sector might be affected, positively or negatively, by climate change and/or climate variability. "Vulnerability" is the degree to which a particular system or sector might be adversely affected by climate change.

The sensitivity of human society to climate change varies. Sectors sensitive to climate change include water resources, coastal zones, human settlements, and human health. Industries sensitive to climate change include agriculture, fisheries, forestry, energy, construction, insurance, financial services, tourism, and recreation.

Food Supply

Net crop production in selected tropical countries and worldwide (2004-6=100)

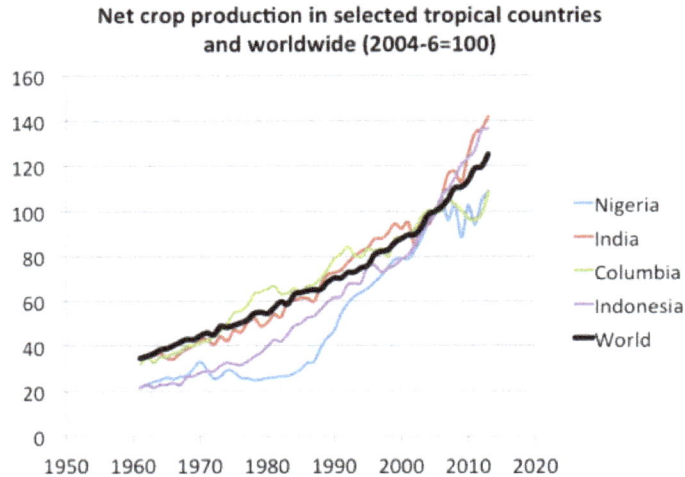

Graph of net crop production worldwide and in selected tropical countries. Raw data from the United Nations.

Climate change will impact agriculture and food production around the world due to: the effects of elevated CO_2 in the atmosphere, higher temperatures, altered precipitation and transpiration regimes, increased frequency of extreme events, and modified weed, pest, and pathogen pressure. In general, low-latitude areas are at most risk of having decreased crop yields.

As of 2007, the effects of regional climate change on agriculture have been small. Changes in crop phenology provide important evidence of the response to recent regional climate change. Phenology is the study of natural phenomena that recur periodically, and how these phenomena relate to climate and seasonal changes. A significant advance in phenology has been observed for agriculture and forestry in large parts of the Northern Hemisphere.

Projections

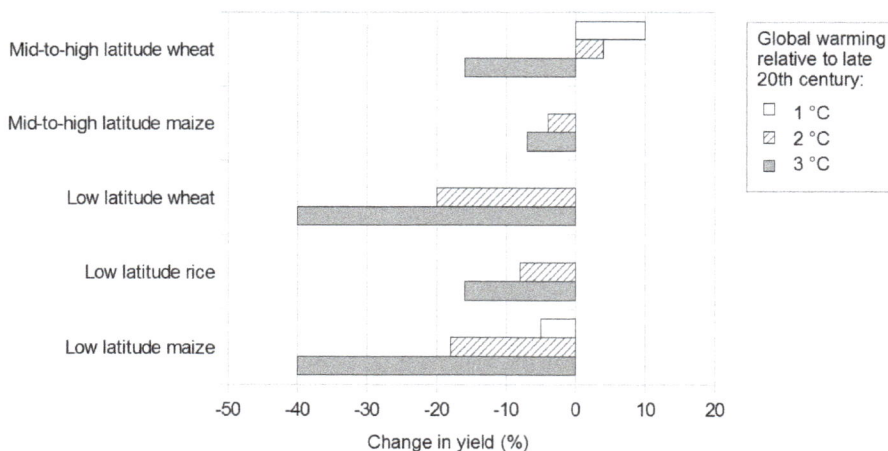

Projected changes in crop yields at different latitudes with global warming. This graph is based on several studies.

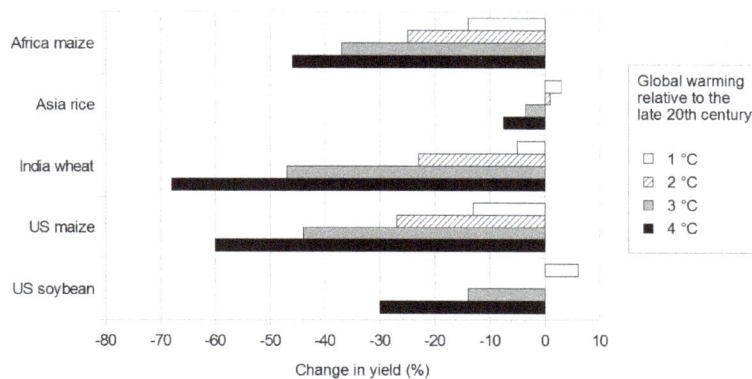

Projected changes in yields of selected crops with global warming. This graph is based on several studies.

With low to medium confidence, Schneider *et al.* (2007) projected that for about a 1 to 3 °C increase in global mean temperature (by the years 2090-2100, relative to average temperatures in the years 1990–2000), there would be productivity decreases for some cereals in low latitudes, and productivity increases in high latitudes. With medium confidence, global production potential was projected to:

- increase up to around 3 °C,

- very likely decrease above about 3 °C.

Most of the studies on global agriculture assessed by Schneider *et al.* (2007) had not incorporated a number of critical factors, including changes in extreme events, or the spread of pests and diseases. Studies had also not considered the development of specific practices or technologies to aid adaptation to climate change.

The graphs opposite show the projected effects of climate change on selected crop yields. Actual changes in yields may be above or below these central estimates.

The projections above can be expressed relative to pre-industrial (1750) temperatures. 0.6 °C of warming is estimated to have occurred between 1750 and 1990-2000. Add 0.6 °C to the above projections to convert them from a 1990-2000 to pre-industrial baseline.

Food Security

Easterling *et al.* (2007) assessed studies that made quantitative projections of climate change impacts on food security. It was noted that these projections were highly uncertain and had limitations. However, the assessed studies suggested a number of fairly robust findings. The first was that climate change would likely increase the number of people at risk of hunger compared with reference scenarios with no climate change. Climate change impacts depended strongly on projected future social and economic development. Additionally, the magnitude of climate change impacts was projected to be smaller compared to the impact of social and economic development. In 2006, the global estimate for the number of people undernourished was 820 million. Under the SRES A1, B1, and B2 scenarios (see the SRES article for information on each scenario group), projections for the year 2080 showed a reduction in the number of people undernourished of about 560-700 million people, with a global total of undernourished people of 100-240 million in 2080.

By contrast, the SRES A2 scenario showed only a small decrease in the risk of hunger from 2006 levels. The smaller reduction under A2 was attributed to the higher projected future population level in this scenario.

Droughts and Agriculture

Some evidence suggests that droughts have been occurring more frequently because of global warming and they are expected to become more frequent and intense in Africa, southern Europe, the Middle East, most of the Americas, Australia, and Southeast Asia. However, other research suggests that there has been little change in drought over the past 60 years. Their impacts are aggravated because of increased water demand, population growth, urban expansion, and environmental protection efforts in many areas. Droughts result in crop failures and the loss of pasture grazing land for livestock.

Health

Human beings are exposed to climate change through changing weather patterns (temperature, precipitation, sea-level rise and more frequent extreme events) and indirectly through changes in water, air and food quality and changes in ecosystems, agriculture, industry and settlements and the economy (Confalonieri *et al.*, 2007:393). According to an assessment of the scientific literature by Confalonieri *et al.* (2007:393), the effects of climate change to date have been small, but are projected to progressively increase in all countries and regions.

A study by the World Health Organization (WHO, 2009) estimated the effect of climate change on human health. Not all of the effects of climate change were included in their estimates, for example, the effects of more frequent and extreme storms were excluded. Climate change was estimated to have been responsible for 3% of diarrhoea, 3% of malaria, and 3.8% of dengue fever deaths worldwide in 2004. Total attributable mortality was about 0.2% of deaths in 2004; of these, 85% were child deaths.

Projections

With high confidence, authors of the IPCC AR4 Synthesis report projected that climate change would bring some benefits in temperate areas, such as fewer deaths from cold exposure, and some mixed effects such as changes in range and transmission potential of malaria in Africa. Benefits were projected to be outweighed by negative health effects of rising temperatures, especially in developing countries.

With very high confidence, Confalonieri *et al.* (2007) concluded that economic development was an important component of possible adaptation to climate change. Economic growth on its own, however, was not judged to be sufficient to insulate the world's population from disease and injury due to climate change. Future vulnerability to climate change will depend not only on the extent of social and economic change, but also on how the benefits and costs of change are distributed in society. For example, in the 19th century, rapid urbanization in western Europe lead to a plummeting in population health. Other factors important in determining the health of populations include education, the availability of health services, and public-health infrastructure.

Water Resources

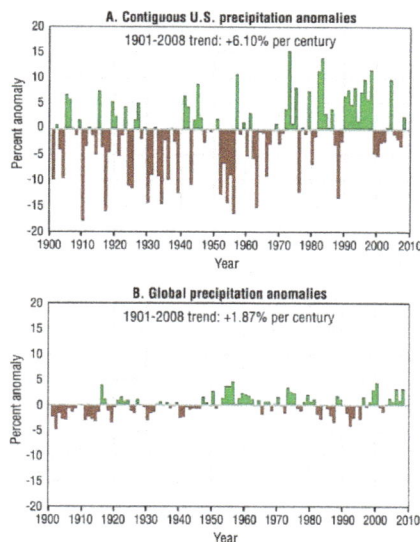

A. Contiguous U.S. precipitation anomalies
1901-2008 trend: +6.10% per century

B. Global precipitation anomalies
1901-2008 trend: +1.87% per century

[a]Anomalies and percent change are calculated with respect to the 1971-2000 mean.
Data source: NOAA, 2009

Precipitation during the 20th century and up through 2008 during global warming, the NOAA estimating an observed trend over that period of 1.87% global precipitation increase per century.

A number of climate-related trends have been observed that affect water resources. These include changes in precipitation, the crysosphere and surface waters (e.g., changes in river flows). Observed and projected impacts of climate change on freshwater systems and their management are mainly due to changes in temperature, sea level and precipitation variability. Sea level rise will extend areas of salinization of groundwater and estuaries, resulting in a decrease in freshwater availability for humans and ecosystems in coastal areas. In an assessment of the scientific literature, Kundzewicz *et al.* (2007) concluded, with high confidence, that:

- the negative impacts of climate change on freshwater systems outweigh the benefits. All of the regions assessed in the IPCC Fourth Assessment Report (Africa, Asia, Australia and New Zealand, Europe, Latin America, North America, Polar regions (Arctic and Antarctic), and small islands) showed an overall net negative impact of climate change on water resources and freshwater ecosystems.

- Semi-arid and arid areas are particularly exposed to the impacts of climate change on freshwater. With very high confidence, it was judged that many of these areas, e.g., the Mediterranean basin, Western United States, Southern Africa, and north-eastern Brazil, would suffer a decrease in water resources due to climate change.

Migration and Conflict

General circulation models project that the future climate change will bring wetter coasts, drier mid-continent areas, and further sea level rise. Such changes could result in the gravest effects of climate change through human migration. Millions might be displaced by shoreline erosions, river and coastal flooding, or severe drought.

Migration related to climate change is likely to be predominantly from rural areas in developing countries to towns and cities. In the short term climate stress is likely to add incrementally to existing migration patterns rather than generating entirely new flows of people.

It has been argued that environmental degradation, loss of access to resources (e.g., water resources), and resulting human migration could become a source of political and even military conflict. Factors other than climate change may, however, be more important in affecting conflict. For example, Wilbanks *et al.* (2007) suggested that major environmentally influenced conflicts in Africa were more to do with the relative abundance of resources, e.g., oil and diamonds, than with resource scarcity. Scott *et al.* (2001) placed only low confidence in predictions of increased conflict due to climate change.

A 2013 study found that significant climatic changes were associated with a higher risk of conflict worldwide, and predicted that "amplified rates of human conflict could represent a large and critical social impact of anthropogenic climate change in both low- and high-income countries." Similarly, a 2014 study found that higher temperatures were associated with a greater likelihood of violent crime, and predicted that global warming would cause millions of such crimes in the United States alone during the 21st century.

Military planners are concerned that global warming is a "threat multiplier". "Whether it is poverty, food and water scarcity, diseases, economic instability, or threat of natural disasters, the broad range of changing climatic conditions may be far reaching. These challenges may threaten stability in much of the world".

Aggregate Impacts

Aggregating impacts adds up the total impact of climate change across sectors and/or regions. Examples of aggregate measures include economic cost (e.g., changes in gross domestic product (GDP) and the social cost of carbon), changes in ecosystems (e.g., changes over land area from one type of vegetation to another), human health impacts, and the number of people affected by climate change. Aggregate measures such as economic cost require researchers to make value judgements over the importance of impacts occurring in different regions and at different times.

Observed Impacts

Global losses reveal rapidly rising costs due to extreme weather-related events since the 1970s. Socio-economic factors have contributed to the observed trend of global losses, e.g., population growth, increased wealth. Part of the growth is also related to regional climatic factors, e.g., changes in precipitation and flooding events. It is difficult to quantify the relative impact of socio-economic factors and climate change on the observed trend. The trend does, however, suggest increasing vulnerability of social systems to climate change.

Projected Impacts

The total economic impacts from climate change are highly uncertain. With medium confidence, Smith *et al.* (2001) concluded that world GDP would change by plus or minus a few percent for a small increase in global mean temperature (up to around 2 °C relative to the 1990 temperature level). Most studies assessed by Smith *et al.* (2001) projected losses in world GDP for a medium increase in global mean temperature (above 2-3 °C relative to the 1990 temperature level), with increasing losses for greater temperature increases. This assessment is consistent with the findings of more recent studies, as reviewed by Hitz and Smith (2004).

Economic impacts are expected to vary regionally. For a medium increase in global mean temperature (2-3 °C of warming, relative to the average temperature between 1990–2000), market sectors in low-latitude and less-developed areas might experience net costs due to climate change. On the other hand, market sectors in high-latitude and developed regions might experience net benefits for this level of warming. A global mean temperature increase above about 2-3 °C (relative to 1990-2000) would very likely result in market sectors across all regions experiencing either declines in net benefits or rises in net costs.

Aggregate impacts have also been quantified in non-economic terms. For example, climate change over the 21st century is likely to adversely affect hundreds of millions of people through increased coastal flooding, reductions in water supplies, increased malnutrition and increased health impacts.

Biological Systems

Observed impacts on Biological Systems

A vast array of physical and biological systems across the Earth are being affected by human-induced global warming.

With very high confidence, Rosenzweig *et al.* (2007) concluded that recent warming had strongly affected natural biological systems. Hundreds of studies have documented responses of ecosystems, plants, and animals to the climate changes that have already occurred. For example, in the Northern Hemisphere, species are almost uniformly moving their ranges northward and up in elevation in search of cooler temperatures. Humans are very likely causing changes in regional temperatures to which plants and animals are responding.

Projected Impacts on Biological Systems

By the year 2100, ecosystems will be exposed to atmospheric CO_2 levels substantially higher than in the past 650,000 years, and global temperatures at least among the highest of those experienced in the past 740,000 years. Significant disruptions of ecosystems are projected to increase with future climate change. Examples of disruptions include disturbances such as fire, drought, pest infestation, invasion of species, storms, and coral bleaching events. The stresses caused by climate change, added to other stresses on ecological systems (e.g., land conversion, land degradation, harvesting, and pollution), threaten substantial damage to or complete loss of some unique ecosystems, and extinction of some critically endangered species.

Climate change has been estimated to be a major driver of biodiversity loss in cool conifer forests, savannas, mediterranean-climate systems, tropical forests, in the Arctic tundra, and in coral reefs. In other ecosystems, land-use change may be a stronger driver of biodiversity loss at least in the near-term. Beyond the year 2050, climate change may be the major driver for biodiversity loss globally.

A literature assessment by Fischlin *et al.* (2007) included a quantitative estimate of the number of species at increased risk of extinction due to climate change. With medium confidence, it was projected that approximately 20 to 30% of plant and animal species assessed so far (in an unbiased sample) would likely be at increasingly high risk of extinction should global mean temperatures exceed a warming of 2 to 3 °C above pre-industrial temperature levels. The uncertainties in this estimate, however, are large: for a rise of about 2 °C the percentage may be as low as 10%, or for about 3 °C, as high as 40%, and depending on biota (all living organisms of an area, the flora and fauna considered as a unit) the range is between 1% and 80%. As global average temperature exceeds 4 °C above pre-industrial levels, model projections suggested that there could be significant extinctions (40-70% of species that were assessed) around the globe.

Assessing whether future changes in ecosystems will be beneficial or detrimental is largely based on how ecosystems are valued by human society. For increases in global average temperature exceeding 1.5 to 2.5 °C (relative to global temperatures over the years 1980-1999) and in concomitant atmospheric CO_2 concentrations, projected changes in ecosystems will have predominantly negative consequences for biodiversity and ecosystems goods and services, e.g., water and food supply.

Abrupt or Irreversible Changes

Physical, ecological and social systems may respond in an abrupt, non-linear or irregular way to climate change. This is as opposed to a smooth or regular response. A quantitative entity behaves "irregularly" when its dynamics are discontinuous (i.e., not smooth), nondifferentiable, unbounded, wildly varying, or otherwise ill-defined. Such behaviour is often termed "singular". Irregular behaviour in Earth systems may give rise to certain thresholds, which, when crossed, may lead to a large change in the system.

Some singularities could potentially lead to severe impacts at regional or global scales. Examples of "large-scale" singularities are discussed in the articles on abrupt climate change, climate change feedback and runaway climate change. It is possible that human-induced climate change could trigger large-scale singularities, but the probabilities of triggering such events are, for the most part, poorly understood.

With low to medium confidence, Smith *et al.* (2001) concluded that a rapid warming of more than 3 °C above 1990 levels would exceed thresholds that would lead to large-scale discontinuities in the climate system. Since the assessment by Smith *et al.* (2001), improved scientific understanding provides more guidance for two large-scale singularities: the role of carbon cycle feedbacks in future climate change (discussed below in the section on biogeochemical cycles) and the melting of the Greenland and West Antarctic ice sheets.

Biogeochemical Cycles

Climate change may have an effect on the carbon cycle in an interactive "feedback" process. A feedback exists where an initial process triggers changes in a second process that in turn influences the initial process. A positive feedback intensifies the original process, and a negative feedback reduces it. Models suggest that the interaction of the climate system and the carbon cycle is one where the feedback effect is positive.

Using the A2 SRES emissions scenario, Schneider *et al.* (2007) found that this effect led to additional warming by the years 2090-2100 (relative to the 1990–2000) of 0.1–1.5 °C. This estimate was made with high confidence. The climate projections made in the IPCC Fourth Assessment Report summarized earlier of 1.1–6.4 °C account for this feedback effect. On the other hand, with medium confidence, Schneider *et al.* (2007) commented that additional releases of GHGs were possible from permafrost, peat lands, wetlands, and large stores of marine hydrates at high latitudes.

Greenland and West Antarctic Ice Sheets

With medium confidence, authors of AR4 concluded that with a global average temperature increase of 1–4 °C (relative to temperatures over the years 1990–2000), at least a partial deglaciation of the Greenland ice sheet, and possibly the West Antarctic ice sheets would occur. The estimated timescale for partial deglaciation was centuries to millennia, and would contribute 4 to 6 metres (13 to 20 ft) or more to sea level rise over this period.

Atlantic Meridional Overturning Circulation

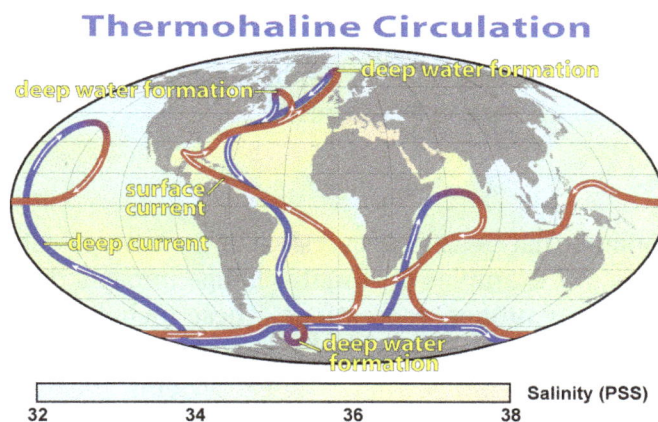

This map shows the general location and direction of the warm surface (red) and cold deep water (blue) currents of the thermohaline circulation. Salinity is represented by color in units of the Practical Salinity Scale. Low values (blue) are less saline, while high values (orange) are more saline.

The Atlantic Meridional Overturning Circulation (AMOC) is an important component of the Earth's climate system, characterized by a northward flow of warm, salty water in the upper layers of the At-lantic and a southward flow of colder water in the deep Atlantic. The AMOC is equivalently known as the thermohaline circulation (THC). Potential impacts associated with MOC changes include reduced warming or (in the case of abrupt change) absolute cooling of northern high-latitude areas near Greenland and north-western Europe, an increased warming of Southern Hemisphere high-lat-itudes, tropical drying, as well as changes to marine ecosystems, terrestrial vegetation, oceanic CO2 uptake, oceanic oxygen concentrations, and shifts in fisheries. According to an assessment by the US Climate Change Science Program (CCSP, 2008b), it is very likely (greater than 90% probability, based on expert judgement) that the strength of the AMOC will decrease over the course of the 21st century. Warming is still expected to occur over most of the European region downstream of the North Atlantic Current in response to increasing GHGs, as well as over North America. Al-though it is very unlikely (less than 10% probability, based on expert judgement) that the AMOC will collapse in the 21st century, the potential consequences of such a collapse could be severe.

Irreversibilities

Commitment to Radiative Forcing

Emissions of GHGs are a potentially irreversible commitment to sustained radiative forcing in the future. The contribution of a GHG to radiative forcing depends on the gas's ability to trap infrared (heat) radiation, the concentration of the gas in the atmosphere, and the length of time the gas resides in the atmosphere.

CO_2 is the most important anthropogenic GHG. While more than half of the CO_2 emitted is currently removed from the atmosphere within a century, some fraction (about 20%) of emitted CO_2 remains in the atmosphere for many thousands of years. Consequently, CO_2 emitted today is potentially an irreversible commitment to sustained radiative forcing over thousands of years.

This commitment may not be truly irreversible should techniques be developed to remove CO_2 or other GHGs directly from the atmosphere, or to block sunlight to induce cooling. Techniques of this sort are referred to as geoengineering. Little is known about the effectiveness, costs or potential side-effects of geoengineering options. Some geoengineering options, such as blocking sunlight, would not prevent further ocean acidification.

Irreversible Impacts

Human-induced climate change may lead to irreversible impacts on physical, biological, and social systems. There are a number of examples of climate change impacts that may be irreversible, at least over the timescale of many human generations. These include the large-scale singularities described above – changes in carbon cycle feedbacks, the melting of the Greenland and West Antarctic ice sheets, and changes to the AMOC. In biological systems, the extinction of species would be an irreversible impact. In social systems, unique cultures may be lost due to climate change. For example, humans living on atoll islands face risks due to sea-level rise, sea-surface warming, and increased frequency and intensity of extreme weather events.

Benefits of Global Warming

With a large range of effects, it is unlikely that all effects will be negative. Not only have some positive effects been identified, but there is some published material indicating that a small amount of warming would be good. The IPCC cautions that "Estimates agree on the size of the impact (small relative to economic growth), and 17 of the 20 impact estimates shown in Figure 10-1 are negative. Losses accelerate with greater warming, and estimates diverge."

The identified benefits are listed below.

CO2 Fertilisation Effect

CO_2 is one of the substances which plants require to grow. Increasing its amount in the air contributes to:

- Improved agriculture in some high latitude regions

- Increased growing season in Greenland

- Increased productivity of sour orange trees

- Increased vegetation activity in high northern latitudes

- Increased plankton biomass in the North Pacific Subtropical Gyre

- Recent increase in forest growth

- Increased Arctic tundra plant reproduction

Human Health

- Winter deaths will decline as temperatures warm

Ice-free Northwest Passage

- Ships will travel on a shorter route between the Pacific and Atlantic oceans

Animal Population Changes

Some animals will benefit from the warming:

- Increase in chinstrap and gentoo penguins

- Bigger marmots

Scientific Opinion

The Intergovernmental Panel on Climate Change (IPCC) has published several major assessments on the effects of global warming. Its most recent comprehensive impact assessment was published in 2014. Publications describing the effects of climate change have also been produced by the following organizations:

- American Association for the Advancement of Science (AAAS)

- A report by the Netherlands Environmental Assessment Agency, the Royal Netherlands Meteorological Institute, and Wageningen University and Research Centre

- UK AVOID research programme

- A report by the UK Royal Society and US National Academy of Sciences

- University of New South Wales Climate Change Research Centre

- US National Research Council

A report by Molina *et al.* (no date) states:

The overwhelming evidence of human-caused climate change documents both current impacts with significant costs and extraordinary future risks to society and natural systems

NASA Data and Tools

NASA has released public data and tools to predict how temperature and rainfall patterns worldwide may change through to the year 2100 caused by increasing carbon dioxide in Earth's atmosphere. The dataset shows projected changes worldwide on a regional level simulated by 21 climate models. The data can be viewed on a daily timescale for individual cities and towns and may be used to conduct climate risk assessments to predict the local and global effects of weather dangers, for example droughts, floods, heat waves and declines in agriculture productivity, and help plan responses to global warming effects.

Effects of Global Warming on Oceans

Waves on an ocean coast

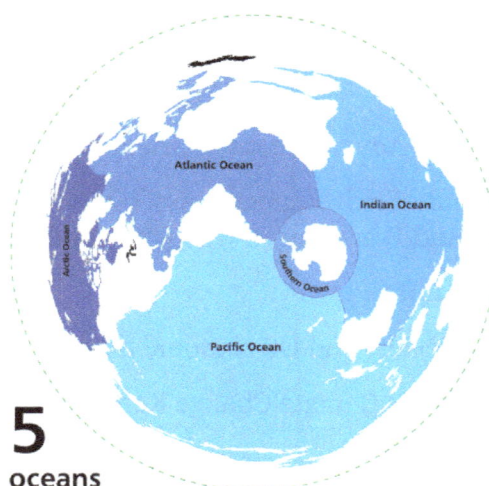

Animated map exhibiting the world's oceanic waters. A continuous body of water encircling the Earth, the World Ocean is divided into a number of principal areas with relatively free interchange among them. Five oceanic divisions are usually reckoned: Pacific, Atlantic, Indian, Arctic, and Southern; the last two listed are sometimes consolidated into the first three.

Global Land–Ocean Temperature Index

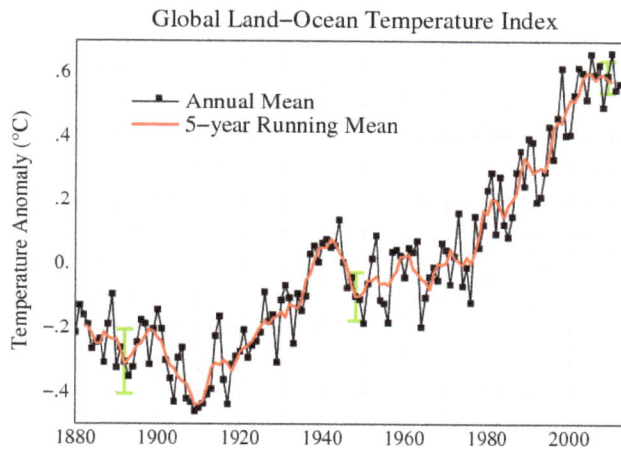

Global mean land-ocean temperature change from 1880–2011, relative to the 1951–1980 mean. The black line is the annual mean and the red line is the 5-year running mean. The green bars show uncertainty estimates. Source: NASA GISS

Where is global warming going?

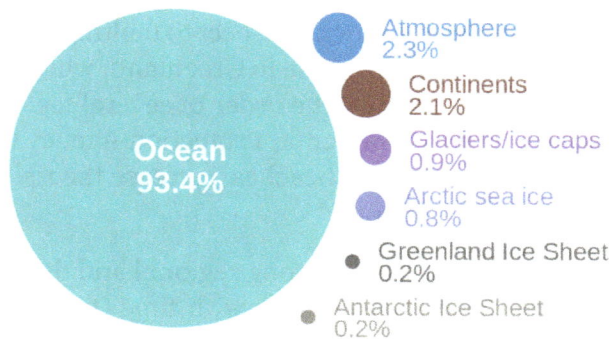

Ocean 93.4%

Atmosphere 2.3%

Continents 2.1%

Glaciers/ice caps 0.9%

Arctic sea ice 0.8%

Greenland Ice Sheet 0.2%

Antarctic Ice Sheet 0.2%

Energy (heat) added to various parts of the climate system due to global warming.

Sea Level

Coasts

There are a number of factors affecting rising sea levels, including the thermal expansion of seawater, the melting of glaciers and ice sheets on land, and possibly human changes to groundwater storage.

The consensus of many studies of coastal tide gauge records is that during the past century sea level has risen worldwide at an average rate of 1–2 mm/yr reflecting a net flux of heat into the surface of the land and oceans. Corresponding studies based on satellite altimetry shows that this rate has increased to closer to 3 mm/yr during the more completely monitored past 20 years. A recent review of the literature suggests that 30% of the sea level rise since 1993 is due to thermal expansion and 55% due to continental ice melt, both resulting from warming global temperatures. In another study, results estimate the heat content of the ocean in the upper 700 meters has increased significantly from 1955-2010. Observations of the changes in heat content of the ocean are important for providing realistic estimates of how the ocean is changing with global warming. An even more recent study of the contributions to global sea level due to melting of the two large ice sheets based on satellite measurements of gravity fluctuations suggests that the melting of these alone are causing global sea level to about 1 mm/yr. In a recent modeling study, scientists used an earth system model to study several variables of the ocean, one of which was the heat content of the oceans over the past several hundred years. The earth system model incorporated the atmosphere, land surface processes, and other earth components to make it more realistic and similar to observations. Results of their model simulation showed that since 1500, the ocean heat content of the upper 500 m has increased.

The connection between sea level rise and ocean thermal expansion follows from Charles's law (also known as the law of volumes) put simply states that the volume of a given mass is proportional to its temperature. This contribution to sea level is monitored by oceanographers using a succession of temperature measuring profiling instruments, which is then compiled at national data centers such as the United States National Oceanographic Data Center. The International Panel on Climate Change (IPCC) Fifth Assessment Report estimates that the upper ocean (surface to 750 m deep) has warmed by 0.09 to 0.13 degrees C per decade over the past 40 years. Other processes important in influencing global sea level include changes to groundwater storage including dams and reservoirs.

Global warming also has an enormous impact with respect to melting glaciers and ice sheets. Higher global temperatures melt glaciers such as the one in Greenland, which flow into the oceans, adding to the amount of seawater. A large rise (on the order of several feet) in global sea levels poses many threats. According to the U.S. Environmental Protection Agency (EPA), "such a rise would inundate coastal wetlands and lowlands, erode beaches, increase the risk of flooding, and increase the salinity of estuaries, aquifers, and wetlands."

Superimposed on the global rise in sea level, is strong regional and decadal variability which may cause sea level along a particular coastline to decline with time (for example along the Canadian eastern seaboard), or to rise faster than the global average. Regions that have shown a rapid rise in sea level during the past two decades include the western tropical Pacific and the United States northeastern seaboard. These regional variations in sea level are the result of many factors, such as local sedimentation rates, geomorphology, post-glacial rebound, and coastal erosion. Large storm events, such as Hurricane Sandy in the eastern Atlantic, can dramatically alter coastlines and affect sea level rise as well.

Coastal regions would be most affected by rising sea levels. The increase in sea level along the coasts of continents, especially North America are much more significant than the global average. According to 2007 estimates by the International Panel on Climate Change (IPCC), "global average sea level will rise between 0.6 and 2 feet (0.18 to 0.59 meters) in the next century. Along the U.S. Mid-Atlantic and Gulf Coasts, however, sea level rose in the last century 5 to 6 inches more than the global average. This is due to the subsiding of coastal lands. The sea level along the U.S. Pacific coast has also increased more than the global average but less than along the Atlantic coast. This can be explained by the varying continental margins along both coasts; the Atlantic type continental margin is characterized by a wide, gently sloping continental shelf, while the Pacific type continental margin incorporates a narrow shelf and slope descending into a deep trench. Since low-sloping coastal regions should retreat faster than higher-sloping regions, the Atlantic coast is more vulnerable to sea level rise than the Pacific coast.

Society

The rise in sea level along coastal regions carries implications for a wide range of habitats and inhabitants. Firstly, rising sea levels will have a serious impact on beaches— a place which humans love to visit recreationally and a prime location for real estate. It is ideal to live on the coast, due to a more moderate climate and pleasant scenery, but beachfront property is at risk from eroding land and rising sea levels. Since the threat posed by rising sea levels has become more prominent, property owners and local government have taken measures to prepare for the worst. For example, "Maine has enacted a policy declaring that shorefront buildings will have to be moved to enable beaches and wetlands to migrate inland to higher ground." Additionally, many coastal states add sand to their beaches to offset shore erosion, and many property owners have elevated their structures in low-lying areas. As a result of the erosion and ruin of properties by large storms on coastal lands, governments have looked into buying land and having residents relocate further inland.

The seas now absorb much of human-generated carbon dioxide, which then affects temperature change. The oceans store 93 percent of that energy which helps keep the planet livable by moderating temperatures.

Another important coastal habitat that is threatened by sea level rise is wetlands, which "occur along the margins of estuaries and other shore areas that are protected from the open ocean and include swamps, tidal flats, coastal marshes and bayous." Wetlands are extremely vulnerable to rising sea levels, since they are within several feet of sea level. The threat posed to wetlands is serious, due to the fact that they are highly productive ecosystems, and they have an enormous impact on the economy of surrounding areas. Wetlands in the U.S. are rapidly disappearing due to an increase in housing, industry, and agriculture, and rising sea levels contribute to this dangerous trend. As a result of rising sea levels, the outer boundaries of wetlands tend to erode, forming new wetlands more inland. According to the EPA, "the amount of newly created wetlands, however, could be much smaller than the lost area of wetlands— especially in developed areas protected with bulkheads, dikes, and other structures that keep new wetlands from forming inland." When estimating a sea level rise within the next century of 50 cm (20 inches), the U.S. would lose 38% to 61% of its existing coastal wetlands.

A rise in sea level will have a negative impact not only on coastal property and economy but on our supply of fresh water. According to the EPA, "Rising sea level increases the salinity of both surface water and ground water through salt water intrusion." Coastal estuaries and aquifers, therefore, are at a high risk of becoming too saline from rising sea levels. With respect to estuaries, an increase in salinity would threaten aquatic animals and plants that cannot tolerate high levels of salinity. Aquifers often serve as a primary water supply to surrounding areas, such as Florida's Biscayne aquifer, which receives freshwater from the Everglades and then supplies water to the Florida Keys. Rising sea levels would submerge low-lying areas of the Everglades, and salinity would greatly increase in portions of the aquifer. The considerable rise in sea level and the decreasing amounts of freshwater along the Atlantic and Gulf coasts would make those areas rather uninhabitable. Many economists predict that global warming will be one of the main economic threats to the West Coast, specifically in California. "Low-lying coastal areas, such as along the Gulf Coast, are particularly vulnerable to sea-level rise and stronger storms—and those risks are reflected in rising insurance rates and premiums. In Florida, for example, the average price of a homeowners' policy increased by 77 percent between 2001 and 2006."

Global issue

Since rising sea levels present a pressing problem not only to coastal communities but to the whole global population as well, much scientific research has been performed to analyze the causes and consequences of a rise in sea level. The U.S. Geological Survey has conducted such research, addressing coastal vulnerability to sea level rise and incorporating six physical variables to analyze the changes in sea level: geomorphology; coastal slope (percent); rate of relative sea level rise (mm/yr); shoreline erosion and acceleration rates (m/yr); mean tidal range (m); and mean wave height (m). The research was conducted on the various coasts of the U.S., and the results are very useful for future reference. Along the Pacific coast, the most vulnerable areas are low-lying beaches, and "their susceptibility is primarily a function of geomorphology and coastal slope." With regard to research performed along the Atlantic coast, the most vulnerable areas to sea level rise were found to be along the Mid-Atlantic coast (Maryland to North Carolina) and Northern Florida, since these are "typically high-energy coastlines where the regional coastal slope is low and where the major landform type is a barrier island." For the Gulf coast, the most vulnerable areas are along the Louisiana-Texas coast. According to the results, "the highest-vulnerability areas are typically lower-lying beach and marsh areas; their susceptibility is primarily a function of geomorphology, coastal slope and rate of relative sea-level rise."

Many humanitarians and environmentalists believe that political policy needs to have a bigger role in carbon dioxide reduction. Humans have a substantial influence on the rise of sea level because we emit increasing levels of carbon dioxide into the atmosphere through automobile use and industry. A higher amount of carbon dioxide in the atmosphere leads to higher global temperatures, which then results in thermal expansion of seawater and melting of glaciers and ice sheets.

Ocean Currents

The currents in the world's oceans are a result of varying temperatures associated with the changing latitudes of our planet. As the atmosphere is warmed nearest the equator, the hot air at the surface of our planet is heated, causing it to rise and draw in cooler air to take its place, creating what is known as circulation cells. This ultimately causes the air to be significantly colder near the poles than at the equator.

Wind patterns associated with these circulation cells drive surface currents which push the surface water to the higher latitudes where the air is colder. This cools the water down enough to where it is capable of dissolving more gasses and minerals, causing it to become very dense in relation to lower latitude waters, which in turn causes it to sink to the bottom of the ocean, forming what is known as North Atlantic Deep Water (NADW) in the north and Antarctic Bottom Water (AABW) in the south. Driven by this sinking and the upwelling that occurs in lower latitudes, as well as the driving force of the winds on surface water, the ocean currents act to circulate water throughout the entire ocean.

When global warming is added into the equation, changes occur, especially in the regions where deep water is formed. With the warming of the oceans and subsequent melting of glaciers and the polar ice caps, more and more fresh water is released into the high latitude regions where deep water is formed. This extra water that gets thrown into the chemical mix dilutes the contents of the water arriving from lower latitudes, reducing the density of the surface water. Consequently the water sinks more slowly than it normally would.

It is important to note that ocean currents provide the necessary nutrients for life to sustain itself in the lower latitudes. Should the currents slow down, fewer nutrients would be brought to sustain ocean life resulting in a crumbling of the food chain and irreparable damage to the marine ecosystem. Slower currents would also mean less carbon fixation. Naturally, the ocean is the largest sink within which carbon is stored. When waters become saturated with carbon, excess carbon has nowhere to go, because the currents are not bringing up enough fresh water to fix the excess. This causes a rise in atmospheric carbon which in turn causes positive feedback that can lead to a runaway greenhouse effect.

Ocean Acidification

Another effect of global warming on the carbon cycle is ocean acidification. The ocean and the atmosphere constantly act to maintain a state of equilibrium, so a rise in atmospheric carbon naturally leads to a rise in oceanic carbon. When carbon is dissolved in water it forms hydrogen and bicarbonate ions, which in turn breaks down to hydrogen and carbonate ions. All these extra hydrogen ions increase the acidity of the ocean and make survival harder for planktonic organisms that depend on calcium carbonate to form their shells. A decrease in the base of the food chain will, once again, be destructive to the ecosystems to which they belong. With fewer of these photosynthetic organisms present at the surface of the ocean, less carbon will be converted to oxygen, thereby allowing the greenhouse gasses to go unchecked.

The effects of ocean acidification can already be seen and have been happening since the start of the industrial revolution, with pH levels of the ocean dropping by 0.1 since the pre-industrial revolution times. An effect called coral bleaching can be seen on the Great Barrier Reef in Australia, where ocean acidification's effects are already taking place. Coral bleaching is when unicellular organisms that help make up the coral begin to die off and leave the coral giving it a white appearance. These unicellular organisms are important for the coral to feed and get the proper nutrition that is necessary to survive, leaving the coral weak and malnourished. This results in weaker coral that can die more easily and offer less protection to the organisms that depend on coral for shelter and protection. Increased acidity can also dissolve an organism's shell, threatening entire groups of shellfish and zooplankton and in turn, presenting a threat to the food chain and ecosystem.

Without strong shells, surviving and growing becomes more of a challenge for marine life that depend on calcified shells. The populations of these animals becomes smaller and individual members of the species turn weaker. The fish that rely on these smaller shell constructing animals for food now have a decreased supply, and animals that need coral reefs for shelter now have less protection. The effects of ocean acidification decrease population sizes of marine life and may cause an economic disruption if enough fish die off, which can seriously harm the global economy as the fishing industry makes a lot of money worldwide.

Steps are being taken to combat the potentially devastating effects of ocean acidification, and scientists worldwide are coming together to solve the problem that is known as "global warming's evil twin".

Marine Life

Research indicates that increasing ocean temperatures are taking a toll on the marine ecosystem. A study on the phytoplankton changes in the Indian Ocean indicates a decline of up to 20% in the marine phytoplankton in the Indian Ocean, during the past six decades. The tuna catch rates have also declined abruptly during the past half century, mostly due to increased industrial fisheries, with the ocean warming adding further stress to the fish species.

Weather

Global warming also affects weather patterns as they pertain to cyclones. Scientists have found that although there have been fewer cyclones than in the past, the intensity of each cyclone has increased. A simplified definition of what global warming means for the planet is that colder regions would get warmer and warmer regions would get much warmer. However, there is also speculation that the complete opposite could be true. A warmer earth could serve to moderate temperatures worldwide. There is still much that is not understood about the earth's climate, because it is very difficult to make climate models. As such, predicting the effects that global warming might have on our planet is still an inexact science.

Sea Floor

Definition

The contents of the ocean floor vary diversely in their origin, from eroded land materials carried into the ocean by rivers or wind flow, waste and decompositions of sea animals, and precipitation of chemicals within the sea water itself, including some from outer space. There are four basic types of sediment of the sea floor:

1.) "Terrigenous" describes the sediment derived from the materials eroded by rain, rivers, glaciers and that which is blown into the ocean by the wind, such as volcanic ash. 2.) Biogenous material is the sediment made up of the hard parts of sea animals that accumulate on the bottom of the ocean. 3.) Hydrogenous sediment is the dissolved material that precipitates in the ocean when oceanic conditions change, and 4.) cosmogenous sediment comes from extraterrestrial sources. These are the components that make up the seafloor under their genetic classifications.

Terrigenous and Biogenous

Terrigenous sediment is the most abundant sediment found on the seafloor, followed by biogenous sediment. The sediment in areas of the ocean floor which is at least 30% biogenous materials is labeled as an ooze. There are two types of oozes: Calcareous oozes and Siliceous oozes. Plankton is the contributor of oozes. Calcareous oozes are predominantly composed of calcium shells found in phytoplankton such as coccolithophores and zooplankton like the foraminiferans. These calcareous oozes are never found deeper than about 4,000 to 5,000 meters because at further depths the calcium dissolves. Similarly, Siliceous oozes are dominated by the siliceous shells of phytoplankton like diatoms and zooplankton such as radiolarians. Depending on the productivity of these planktonic organisms, the shell material that collects when these organisms die may build up at a rate anywhere from 1mm to 1 cm every 1000 years.

Hydrogenous and Cosmogenous

Hydrogenous sediments are uncommon. They only occur with changes in oceanic conditions such as temperature and pressure. Rarer still are cosmogenous sediments. Hydrogenous sediments are formed from dissolved chemicals that precipitate from the ocean water, or along the mid-ocean ridges, they can form by metallic elements binding onto rocks that have water of more than 300 degrees Celsius circulating around them. When these elements mix with the cold sea water they precipitate from the cooling water. Known as manganese nodules, they are composed of layers of different metals like manganese, iron, nickel, cobalt, and copper, and they are always found on the surface of the ocean floor. Cosmogenous sediments are the remains of space debris such as comets and asteroids, made up of silicates and various metals that have impacted the Earth.

Size Classification

Another way that sediments are described is through their descriptive classification. These sediments vary in size, anywhere from 1/4096 of a mm to greater than 256 mm. The different types are: boulder, cobble, pebble, granule, sand, silt, and clay, each type becoming finer in grain. The grain size indicates the type of sediment and the environment in which it was created. Larger grains sink faster and can only be pushed by rapid flowing water (high energy environment) whereas small grains sink very slowly and can be suspended by slight water movement, accumulating in conditions where water is not moving so quickly. This means that larger grains of sediment may come together in higher energy conditions and smaller grains in lower energy conditions.

Various amounts of these sediments are deposited around the world and are distributed in three ways: by the processes of production, dilution, and destruction.

Climate Change

It is known that climate affects the ocean and the ocean affects the climate. Due to climate change, as the ocean gets warmer this too has an effect on the seafloor. Because of greenhouse gases such as carbon dioxide, this warming will have an effect on the bicarbonate buffer of the ocean. The bicarbonate buffer is the concentration of bicarbonate ions that keeps the ocean's acidity balanced within a pH range of 7.5-8.4. Addition of carbon dioxide to the ocean water makes the oceans more acidic. Increased ocean acidity is not good for the planktonic organisms that depend on calcium to form their shells. Calcium dissolves with very weak acids and any increase in the ocean's acidity will be destructive for the calcareous organisms. Increased ocean acidity will lead to decreased Calcite Compensation Depth (CCD), causing calcite to dissolve in shallower waters. This will then have a great effect on the calcareous ooze in the ocean, because the sediment itself would begin to dissolve.

Predictions

If ocean temperatures rise it will have an effect right beneath the ocean floor and it will allow the addition of another greenhouse gas, methane gas. Methane gas has been found under methane hydrate, frozen methane and water, beneath the ocean floor. With the ocean warming, this methane hydrate will begin to melt and release methane gas, contributing to global warming. Increase of water temperature will also have a devastating effect on different oceanic ecosystems like coral reefs.

The direct effect is the coral bleaching of these reefs, which live within a narrow temperature margin, so a small increase in temperature would have a drastic effects in these environments. When corals bleach it is because the coral loses 60-90% of their zooxanthellae due to various stressors, ocean temperature being one of them. If the bleaching is prolonged, the coral host would die.

Global Warming Projections

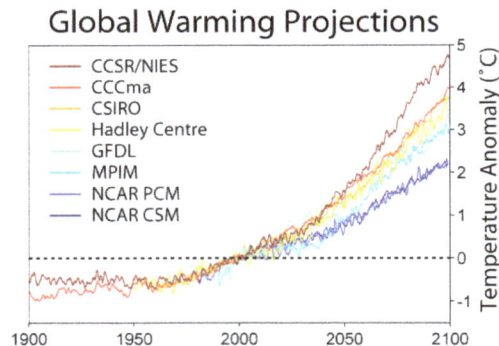

Calculations prepared in or before 2001 from a range of climate models under the SRES A2 emissions scenario, which assumes no action is taken to reduce emissions and regionally divided economic development.

Global Warming Predictions

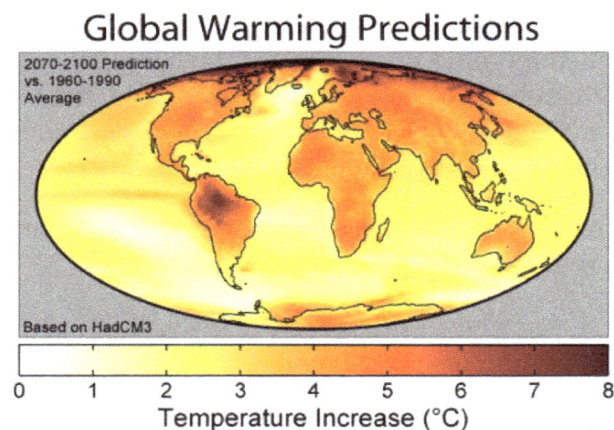

The geographic distribution of surface warming during the 21st century calculated by the HadCM3 climate model if a business as usual scenario is assumed for economic growth and greenhouse gas emissions. In this figure, the globally averaged warming corresponds to 3.0 °C (5.4 °F).

Although uncertain, another effect of climate change may be the growth, toxicity, and distribution of harmful algal blooms. These algal blooms have serious effects on not only marine ecosystems, killing sea animals and fish with their toxins, but also for humans as well. Some of these blooms deplete the oxygen around them to levels low enough to kill fish. It is important that these harmful effects be noted with higher levels of awareness, so that changes can be implemented before it's too late.

Ocean Acidification

Ocean acidification is the ongoing decrease in the pH of the Earth's oceans, caused by the uptake of carbon dioxide (CO_2) from the atmosphere. Seawater is slightly basic (meaning pH > 7), and the process in question is a shift towards pH-neutral conditions rather than a transition to acidic conditions (pH < 7). Ocean alkalinity is not changed by the process, or may increase over long time periods due to carbonate dissolution. An estimated 30–40% of the carbon dioxide from human activity released into the atmosphere dissolves into oceans, rivers and lakes. To achieve chemical equilibrium, some of it reacts with the water to form carbonic acid. Some of these extra carbonic acid molecules react with a water molecule to give a bicarbonate ion and a hydronium ion, thus increasing ocean acidity (H^+ ion concentration). Between 1751 and 1994 surface ocean pH is estimated to have decreased from approximately 8.25 to 8.14, representing an increase of almost 30% in H^+ ion concentration in the world's oceans. Earth System Models project that within the last decade ocean acidity exceeded historical analogs and in combination with other ocean biogeochemical changes could undermine the functioning of marine ecosystems and disrupt the provision of many goods and services associated with the ocean.

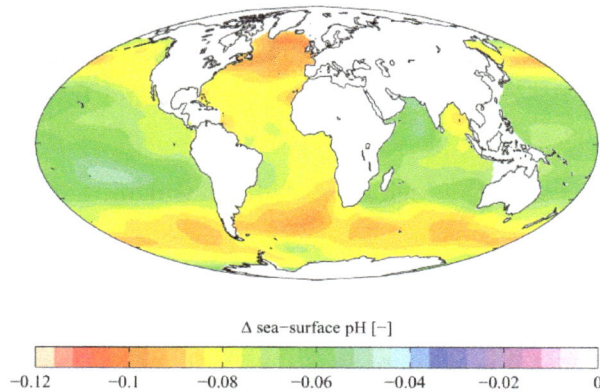

Estimated change in sea water pH caused by human created CO2 between the 1700s and the 1990s, from the Global Ocean Data Analysis Project (GLODAP) and the World Ocean Atlas

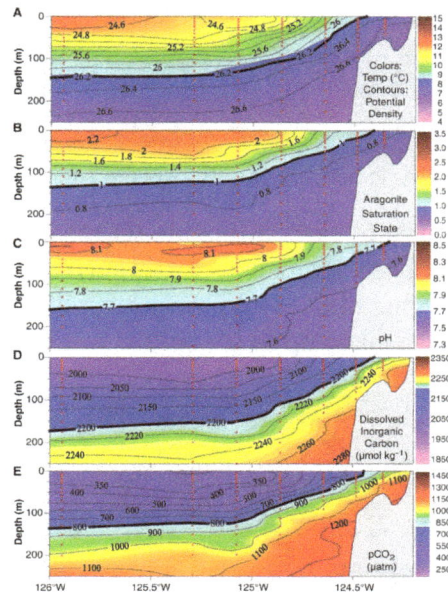

NOAA provides evidence for upwelling of "acidified" water onto the Continental Shelf. In the figure above, note the vertical sections of (A) temperature, (B) aragonite saturation, (C) pH, (D) DIC, and (E) pCO2 on transect line 5 off Pt. St. George, California. The potential density surfaces are superimposed on the temperature section. The 26.2 potential density surface delineates the location of the first instance in which the undersaturated water is upwelled from depths of 150 to 200 m onto the shelf and outcropping at the surface near the coast. The red dots represent sample locations.

Increasing acidity is thought to have a range of potentially harmful consequences for marine organisms, such as depressing metabolic rates and immune responses in some organisms, and causing coral bleaching. By increasing the presence of free hydrogen ions, each molecule of carbonic acid that forms in the oceans ultimately results in the conversion of *two* carbonate ions into bicarbonate ions. This net decrease in the amount of carbonate ions available makes it more difficult for marine calcifying organisms, such as coral and some plankton, to form biogenic calcium carbonate, and such structures become vulnerable to dissolution. Ongoing acidification of the oceans threatens food chains connected with the oceans. As members of the InterAcademy Panel, 105 science academies have issued a statement on ocean acidification recommending that by 2050, global CO_2 emissions be reduced by at least 50% compared to the 1990 level.

While ongoing ocean acidification is anthropogenic in origin, it has occurred previously in Earth's history. The most notable example is the Paleocene-Eocene Thermal Maximum (PETM), which occurred approximately 56 million years ago. For reasons that are currently uncertain, massive amounts of carbon entered the ocean and atmosphere, and led to the dissolution of carbonate sediments in all ocean basins.

Ocean acidification has been called the "evil twin of global warming" and "the other CO_2 problem".

Carbon cycle

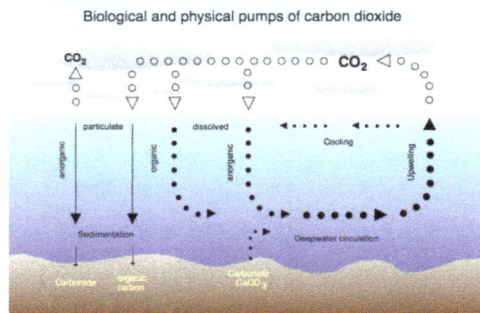

The CO2 cycle between the atmosphere and the ocean

The carbon cycle describes the fluxes of carbon dioxide (CO2) between the oceans, terrestrial biosphere, lithosphere, and the atmosphere. Human activities such as the combustion of fossil fuels and land use changes have led to a new flux of CO2 into the atmosphere. About 45% has remained in the atmosphere; most of the rest has been taken up by the oceans, with some taken up by terrestrial plants.

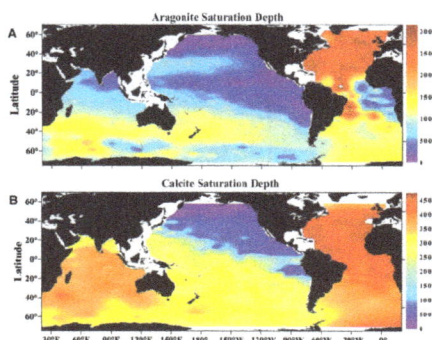

Distribution of (A) aragonite and (B) calcite saturation depth in the global oceans

Changes in Aragonite Saturation of the World's Oceans, 1880–2012

Change in aragonite saturation at the ocean surface (Ω_{ar}):

| -0.8 | -0.7 | -0.6 | -0.5 | -0.4 | -0.3 | -0.2 | -0.1 | 0 |

Data source: Feely, R.A., S.C. Doney, and S.R. Cooley. 2009. Ocean acidification: Present conditions and future changes in a high-CO_2 world. Oceanography 22(4):36–47.

For more information, visit U.S. EPA's "Climate Change Indicators in the United States" at www.epa.gov/climatechange/indicators.

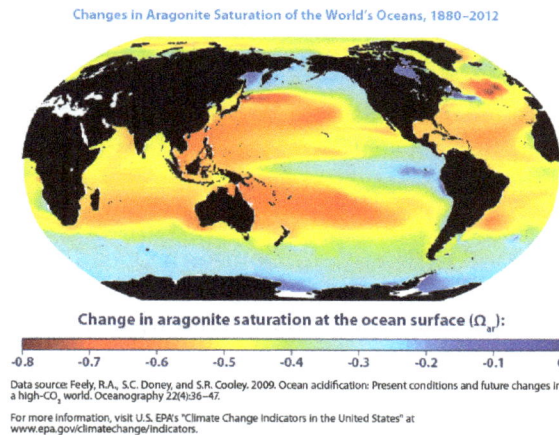

The map was created by the National Oceanic and Atmospheric Administration and the Woods Hole Oceanographic Institution using Community Earth System Model data. This map was created by comparing average conditions during the 1880s with average conditions during the most recent 10 years (2003–2012). Aragonite saturation has only been measured at selected locations during the last few decades, but it can be calculated reliably for different times and locations based on the relationships scientists have observed among aragonite saturation, pH, dissolved carbon, water temperature, concentrations of carbon dioxide in the atmosphere, and other factors that can be measured. This map shows changes in the amount of aragonite dissolved in ocean surface waters between the 1880s and the most recent decade (2003–2012). Aragonite saturation is a ratio that compares the amount of aragonite that is actually present with the total amount of aragonite that the water could hold if it were completely saturated. The more negative the change in aragonite saturation, the larger the decrease in aragonite available in the water, and the harder it is for marine creatures to produce their skeletons and shells. The global map shows changes over time in the amount of aragonite dissolved in ocean water, which is called aragonite saturation.

The carbon cycle involves both organic compounds such as cellulose and inorganic carbon compounds such as carbon dioxide and the carbonates. The inorganic compounds are particularly relevant when discussing ocean acidification for it includes many forms of dissolved CO 2 present in the Earth's oceans.

When CO2 dissolves, it reacts with water to form a balance of ionic and non-ionic chemical species: dissolved free carbon dioxide (CO2(aq)), carbonic acid (H2CO3), bicarbonate (HCO−3) and carbonate (CO2−3). The ratio of these species depends on factors such as seawater temperature and alkalinity (as shown in a Bjerrum plot). These different forms of dissolved inorganic carbon are transferred from an ocean's surface to its interior by the ocean's solubility pump.

The resistance of an area of ocean to absorbing atmospheric CO2 is known as the Revelle factor.

Acidification

Dissolving CO2 in seawater increases the hydrogen ion (H+) concentration in the ocean, and thus decreases ocean pH, as follows:

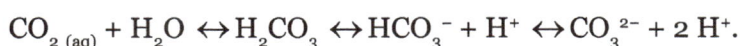

$$CO_{2\ (aq)} + H_2O \leftrightarrow H_2CO_3 \leftrightarrow HCO_3^- + H^+ \leftrightarrow CO_3^{2-} + 2\ H^+.$$

Caldeira and Wickett (2003) placed the rate and magnitude of modern ocean acidification changes in the context of probable historical changes during the last 300 million years.

Since the industrial revolution began, it is estimated that surface ocean pH has dropped by slightly more than 0.1 units on the logarithmic scale of pH, representing about a 29% increase in H+. It is expected to drop by a further 0.3 to 0.5 pH units (an additional doubling to tripling of today's post-industrial acid concentrations) by 2100 as the oceans absorb more anthropogenic CO_2, the impacts being most severe for coral reefs and the Southern Ocean. These changes are predicted to continue rapidly as the oceans take up more anthropogenic CO_2 from the atmosphere. The degree of change to ocean chemistry, including ocean pH, will depend on the mitigation and emissions pathways society takes.

Although the largest changes are expected in the future, a report from NOAA scientists found large quantities of water undersaturated in aragonite are already upwelling close to the Pacific continental shelf area of North America. Continental shelves play an important role in marine ecosystems since most marine organisms live or are spawned there, and though the study only dealt with the area from Vancouver to Northern California, the authors suggest that other shelf areas may be experiencing similar effects.

Average surface ocean pH				
Time	pH	pH change relative to pre-industrial	Source	H+ concentration change relative to pre-industrial
Pre-industrial (18th century)	8.179		analysed field	
Recent past (1990s)	8.104	−0.075	field	+ 18.9%
Present levels	~8.069	−0.11	field	**+ 28.8%**
2050 (2×CO2 = 560 ppm)	7.949	−0.230	model	+ 69.8%
2100 (IS92a)	7.824	−0.355	model	+ 126.5%

Rate

One of the first detailed datasets to examine how pH varied over a period of time at a temperate coastal location found that acidification was occurring much faster than previously predicted, with consequences for near-shore benthic ecosystems. Thomas Lovejoy, former chief biodiversity advisor to the World Bank, has suggested that "the acidity of the oceans will more than double in the next 40 years. This rate is 100 times faster than any changes in ocean acidity in the last 20 million years, making it unlikely that marine life can somehow adapt to the changes." It is predicted that, by the year 2100, the level of acidity in the ocean will reach the levels experienced by the earth 20 million years ago.

Current rates of ocean acidification have been compared with the greenhouse event at the Paleocene–Eocene boundary (about 55 million years ago) when surface ocean temperatures rose by 5–6 degrees Celsius. No catastrophe was seen in surface ecosystems, yet bottom-dwelling organisms in the deep ocean experienced a major extinction. The current acidification is on a path to reach levels higher than any seen in the last 65 million years, and the rate of increase is about ten times the rate that preceded the Paleocene–Eocene mass extinction. The current and projected acidification has been described as an almost unprecedented geological event. A National Research Council study released in April 2010 likewise concluded that "the level of acid in the oceans is increasing at an unprecedented rate." A 2012 paper in the journal *Science* examined the geological record in an attempt to find a historical analog for current global conditions as well as those of the future. The researchers determined that the current rate of ocean acidification is faster than at any time in the past 300 million years.

A review by climate scientists at the RealClimate blog, of a 2005 report by the Royal Society of the UK similarly highlighted the centrality of the *rates* of change in the present anthropogenic acidification process, writing:

"The natural pH of the ocean is determined by a need to balance the deposition and burial of $CaCO_3$ on the sea floor against the influx of $Ca2+$ and $CO2-3$ into the ocean from dissolving rocks on land, called weathering. These processes stabilize the pH of the ocean, by a mechanism called $CaCO_3$ compensation...The point of bringing it up again is to note that if the CO_2 concentration of the atmosphere changes more slowly than this, as it always has throughout the Vostok record, the pH of the ocean will be relatively unaffected because $CaCO_3$ compensation can keep up. The [present] fossil fuel acidification is much faster than natural changes, and so the acid spike will be more intense than the earth has seen in at least 800,000 years."

In the 15-year period 1995–2010 alone, acidity has increased 6 percent in the upper 100 meters of the Pacific Ocean from Hawaii to Alaska. According to a statement in July 2012 by Jane Lubchenco, head of the U.S. National Oceanic and Atmospheric Administration "surface waters are changing much more rapidly than initial calculations have suggested. It's yet another reason to be very seriously concerned about the amount of carbon dioxide that is in the atmosphere now and the additional amount we continue to put out."

A 2013 study claimed acidity was increasing at a rate 10 times faster than in any of the evolutionary crises in Earth's history. In a synthesis report published in *Science* in 2015, 22 leading marine scientists stated that CO_2 from burning fossil fuels is changing the oceans' chemistry more rapidly than at any time since the Great Dying, Earth's most severe known extinction event, emphasizing that the 2 °C maximum temperature increase agreed upon by governments reflects too small a cut in emissions to prevent "dramatic impacts" on the world's oceans, with lead author Jean-Pierre Gattuso remarking that "The ocean has been minimally considered at previous climate negotiations. Our study provides compelling arguments for a radical change at the UN conference (in Paris) on climate change".

Calcification

Overview

Changes in ocean chemistry can have extensive direct and indirect effects on organisms and their habitats. One of the most important repercussions of increasing ocean acidity relates to the production of shells and plates out of calcium carbonate ($CaCO_3$). This process is called calcification and is important to the biology and survival of a wide range of marine organisms. Calcification involves the precipitation of dissolved ions into solid CaCO3 structures, such as coccoliths. After they are formed, such structures are vulnerable to dissolution unless the surrounding seawater contains saturating concentrations of carbonate ions (CO_3^{2-}).

Mechanism

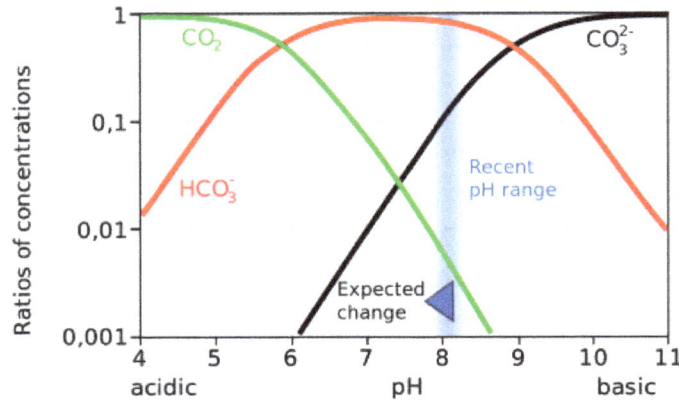

Bjerrum plot: Change in carbonate system of seawater from ocean acidification.

Of the extra carbon dioxide added into the oceans, some remains as dissolved carbon dioxide, while the rest contributes towards making additional bicarbonate (and additional carbonic acid). This also increases the concentration of hydrogen ions, and the percentage increase in hydrogen is larger than the percentage increase in bicarbonate, creating an imbalance in the reaction HCO_3^- $\leftrightarrow CO_3^{2-} + H^+$. To maintain chemical equilibrium, some of the carbonate ions already in the ocean combine with some of the hydrogen ions to make further bicarbonate. Thus the ocean's concentration of carbonate ions is reduced, creating an imbalance in the reaction $Ca^{2+} + CO_3^{2-} \leftrightarrow CaCO_3$, and making the dissolution of formed CaCO3 structures more likely.

These increases in concentrations of dissolved carbon dioxide and bicarbonate, and reduction in carbonate, are shown in a Bjerrum plot.

Saturation State

The saturation state (known as Ω) of seawater for a mineral is a measure of the thermodynamic potential for the mineral to form or to dissolve, and is described by the following equation:

$$\Omega = \frac{\left[Ca^{2+} \right]\left[CO_3^{2-} \right]}{K_{sp}}$$

Here Ω is the product of the concentrations (or activities) of the reacting ions that form the mineral (Ca2+ and CO2−3), divided by the product of the concentrations of those ions when the mineral is at equilibrium (Ksp), that is, when the mineral is neither forming nor dissolving. In seawater, a natural horizontal boundary is formed as a result of temperature, pressure, and depth, and is known as the saturation horizon, or lysocline. Above this saturation horizon, Ω has a value greater than 1, and CaCO3 does not readily dissolve. Most calcifying organisms live in such waters. Below this depth, Ω has a value less than 1, and CaCO3 will dissolve. However, if its production rate is high enough to offset dissolution, CaCO3 can still occur where Ω is less than 1. The carbonate compensation depth occurs at the depth in the ocean where production is exceeded by dissolution.

The decrease in the concentration of CO_3^{2-} decreases Ω, and hence makes CaCO3 dissolution more likely.

Calcium carbonate occurs in two common polymorphs (crystalline forms): aragonite and calcite. Aragonite is much more soluble than calcite, so the aragonite saturation horizon is always nearer to the surface than the calcite saturation horizon. This also means that those organisms that produce aragonite may be more vulnerable to changes in ocean acidity than those that produce calcite. Increasing CO2 levels and the resulting lower pH of seawater decreases the saturation state of CaCO3 and raises the saturation horizons of both forms closer to the surface. This decrease in saturation state is believed to be one of the main factors leading to decreased calcification in marine organisms, as the inorganic precipitation of CaCO3 is directly proportional to its saturation state.

Possible Impacts

Video summarizing the impacts of ocean acidification. Source: NOAA Environmental Visualization Laboratory.

Increasing acidity has possibly harmful consequences, such as depressing metabolic rates in jumbo squid, depressing the immune responses of blue mussels, and coral bleaching. However it may benefit some species, for example increasing the growth rate of the sea star, *Pisaster ochraceus*, while shelled plankton species may flourish in altered oceans.

The report "Ocean Acidification Summary for Policymakers 2013" describes research findings and possible impacts.

Impacts on Oceanic Calcifying Organisms

Although the natural absorption of CO2 by the world's oceans helps mitigate the climatic effects of anthropogenic emissions of CO2, it is believed that the resulting decrease in pH will have negative consequences, primarily for oceanic calcifying organisms. These span the food chain from autotrophs to heterotrophs and include organisms such as coccolithophores, corals, foraminifera, echinoderms, crustaceans and molluscs. As described above, under normal conditions, calcite and aragonite are stable in surface waters since the carbonate ion is at supersaturating concentrations. However, as ocean pH falls, the concentration of carbonate ions required for saturation to occur increases, and when carbonate becomes undersaturated, structures made of calcium carbonate are vulnerable to dissolution. Therefore, even if there is no change in the rate of calcification, the rate of dissolution of calcareous material increases.

Corals, coccolithophore algae, coralline algae, foraminifera, shellfish and pteropods experience reduced calcification or enhanced dissolution when exposed to elevated CO2.

The Royal Society published a comprehensive overview of ocean acidification, and its potential consequences, in June 2005. However, some studies have found different response to ocean acidification, with coccolithophore calcification and photosynthesis both increasing under elevated atmospheric pCO_2, an equal decline in primary production and calcification in response to elevated CO_2 or the direction of the response varying between species. A study in 2008 examining a sediment core from the North Atlantic found that while the species composition of coccolithophorids has remained unchanged for the industrial period 1780 to 2004, the calcification of coccoliths has increased by up to 40% during the same time. A 2010 study from Stony Brook University suggested that while some areas are overharvested and other fishing grounds are being restored, because of ocean acidification it may be impossible to bring back many previous shellfish populations. While the full ecological consequences of these changes in calcification are still uncertain, it appears likely that many calcifying species will be adversely affected.

When exposed in experiments to pH reduced by 0.2 to 0.4, larvae of a temperate brittlestar, a relative of the common sea star, fewer than 0.1 percent survived more than eight days. There is also a suggestion that a decline in the coccolithophores may have secondary effects on climate, contributing to global warming by decreasing the Earth's albedo via their effects on oceanic cloud cover. All marine ecosystems on Earth will be exposed to changes in acidification and several other ocean biogeochemical changes.

The fluid in the internal compartments where corals grow their exoskeleton is also extremely important for calcification growth. When the saturation rate of aragonite in the external seawater is at ambient levels, the corals will grow their aragonite crystals rapidly in their internal compartments, hence their exoskeleton grows rapidly. If the level of aragonite in the external seawater is lower than the ambient level, the corals have to work harder to maintain the right balance in the internal compartment. When that happens, the process of growing the crystals slows down, and this slows down the rate of how much their exoskeleton is growing. Depending on how much aragonite is in the surrounding water, the corals may even stop growing because the levels of aragonite are too low to pump in to the internal compartment. They could even dissolve faster than they can make the crystals to their skeleton, depending on the aragonite levels in the surrounding water.

Ocean acidification may force some organisms to reallocate resources away from productive endpoints such as growth in order to maintain calcification.

In some places carbon dioxide bubbles out from the sea floor, locally changing the pH and other aspects of the chemistry of the seawater. Studies of these carbon dioxide seeps have documented a variety of responses by different organisms. Coral reef communities located near carbon dioxide seeps are of particular interest because of the sensitivity of some corals species to acidification. In Papua New Guinea, declining pH caused by carbon dioxide seeps is associated with declines in coral species diversity. However, in Palau carbon dioxide seeps are not associated with reduced species diversity of corals, although bioerosion of coral skeletons is much higher at low pH sites.

Other Biological Impacts

Aside from the slowing and/or reversing of calcification, organisms may suffer other adverse effects, either indirectly through negative impacts on food resources, or directly as reproductive or physiological effects. For example, the elevated oceanic levels of CO_2 may produce CO_2-induced acidification of body fluids, known as hypercapnia. Also, increasing ocean acidity is believed to have a range of direct consequences. For example, increasing acidity has been observed to: reduce metabolic rates in jumbo squid; depress the immune responses of blue mussels; and make it harder for juvenile clownfish to tell apart the smells of non-predators and predators, or hear the sounds of their predators. This is possibly because ocean acidification may alter the acoustic properties of seawater, allowing sound to propagate further, and increasing ocean noise. Calcium carbonate ions are very important when it comes to building organisms as they use calcium to build skeletons and shells. OA affects ocean species to a varying degree. Many plants do well with high CO_2 levels, but organisms that calcify like clams, mussels, corals and sea urchins might not do so well because everything in the ocean is connected including the food web. Since OA is such a large event because the ocean takes up 71% of the earth's surface OA is being referred to as climate change because it will affect the whole planet if the ocean becomes acidic.[contradictory] This impacts all animals that use sound for echolocation or communication. Atlantic longfin squid eggs took longer to hatch in acidified water, and the squid's statolith was smaller and malformed in animals placed in sea water with a lower pH. The lower PH was simulated with 20-30 times the normal amount of CO_2. However, as with calcification, as yet there is not a full understanding of these processes in marine organisms or ecosystems.

Another possible effect would be an increase in red tide events, which could contribute to the accumulation of toxins (domoic acid, brevetoxin, saxitoxin) in small organisms such as anchovies and shellfish, in turn increasing occurrences of amnesic shellfish poisoning, neurotoxic shellfish poisoning and paralytic shellfish poisoning.

Nonbiological Impacts

Leaving aside direct biological effects, it is expected that ocean acidification in the future will lead to a significant decrease in the burial of carbonate sediments for several centuries, and even the dissolution of existing carbonate sediments. This will cause an elevation of ocean alkalinity, leading to the enhancement of the ocean as a reservoir for CO_2 with implications for climate change as more CO_2 leaves the atmosphere for the ocean.

Impact on Human Industry

The threat of acidification includes a decline in commercial fisheries and in the Arctic tourism industry and economy. Commercial fisheries are threatened because acidification harms calcifying organisms which form the base of the Arctic food webs.

Pteropods and brittle stars both form the base of the Arctic food webs and are both seriously damaged from acidification. Pteropods shells dissolve with increasing acidification and the brittle stars lose muscle mass when re-growing appendages. For pteropods to create shells they require aragonite which is produced through carbonate ions and dissolved calcium. Pteropods are severely affected because increasing acidification levels have steadily decreased the amount of water supersaturated with carbonate which is needed for aragonite creation. Arctic waters are changing so rapidly that they will become undersaturated with aragonite as early as 2016. Additionally the brittle star's eggs die within a few days when exposed to expected conditions resulting from Arctic acidification. Acidification threatens to destroy Arctic food webs from the base up. Arctic food webs are considered simple, meaning there are few steps in the food chain from small organisms to larger predators. For example, pteropods are "a key prey item of a number of higher predators - larger plankton, fish, seabirds, whales" Both pteropods and sea stars serve as a substantial food source and their removal from the simple food web would pose a serious threat to the whole ecosystem. The effects on the calcifying organisms at the base of the food webs could potentially destroy fisheries. The value of fish caught from US commercial fisheries in 2007 was valued at $3.8 billion and of that 73% was derived from calcifiers and their direct predators. Other organisms are directly harmed as a result of acidification. For example, decrease in the growth of marine calcifiers such as the American Lobster, Ocean Quahog, and scallops means there is less shellfish meat available for sale and consumption. Red king crab fisheries are also at a serious threat because crabs are calcifiers and rely on carbonate ions for shell development. Baby red king crab when exposed to increased acidification levels experienced 100% mortality after 95 days. In 2006 Red King Cab accounted for 23% of the total guideline harvest levels and a serious decline in red crab population would threaten the crab harvesting industry. Several ocean goods and services are likely to be undermined by future ocean acidification potentially affecting the livelihoods of some 400 to 800 million people depending upon the emission scenario.

Impact on Indigenous Peoples

Acidification could damage the Arctic tourism economy and affect the way of life of indigenous peoples. A major pillar of Arctic tourism is the sport fishing and hunting industry. The sport fishing industry is threatened by collapsing food webs which provide food for the prized fish. A decline in tourism lowers revenue input in the area, and threatens the economies that are increasingly dependent on tourism. Acidification is not merely a threat but has significantly declined whole fish populations. For example, In Scandinavia studies conducted on acidic water revealed that 15% of species populations had disappeared and that many more populations were limited in numbers or declining. The rapid decrease or disappearance of marine life could also affect the diet of Indigenous peoples.

Possible Responses

Reducing CO_2 Emissions

Members of the InterAcademy Panel recommended that by 2050, global anthropogenic CO_2 emissions be reduced less than 50% of the 1990 level. The 2009 statement also called on world leaders to:

- Acknowledge that ocean acidification is a direct and real consequence of increasing atmospheric CO_2 concentrations, is already having an effect at current concentrations, and is likely to cause grave harm to important marine ecosystems as CO_2 concentrations reach 450 [parts-per-million (ppm)] and above;

- [...] Recognise that reducing the build up of CO_2 in the atmosphere is the only practicable solution to mitigating ocean acidification;

- [...] Reinvigorate action to reduce stressors, such as overfishing and pollution, on marine ecosystems to increase resilience to ocean acidification.

Stabilizing atmospheric CO_2 concentrations at 450 ppm would require near-term emissions reductions, with steeper reductions over time.

The German Advisory Council on Global Change stated:

In order to prevent disruption of the calcification of marine organisms and the resultant risk of fundamentally altering marine food webs, the following guard rail should be obeyed: the pH of near surface waters should not drop more than 0.2 units below the pre-industrial average value in any larger ocean region (nor in the global mean).

One policy target related to ocean acidity is the magnitude of future global warming. Parties to the United Nations Framework Convention on Climate Change (UNFCCC) adopted a target of limiting warming to below 2 °C, relative to the pre-industrial level. Meeting this target would require substantial reductions in anthropogenic CO_2 emissions.

Limiting global warming to below 2 °C would imply a reduction in surface ocean pH of 0.16 from pre-industrial levels. This would represent a substantial decline in surface ocean pH.

Climate Engineering

Climate engineering (mitigating temperature or pH effects of emissions) has been proposed as a possible response to ocean acidification. The IAP (2009) statement cautioned against climate engineering as a policy response:

Mitigation approaches such as adding chemicals to counter the effects of acidification are likely to be expensive, only partly effective and only at a very local scale, and may pose additional unanticipated risks to the marine environment. There has been very little research on the feasibility and impacts of these approaches. Substantial research is needed before these techniques could be applied.

Reports by the WGBU (2006), the UK's Royal Society (2009), and the US National Research Council (2011) warned of the potential risks and difficulties associated with climate engineering.

Iron fertilization

Iron fertilization of the ocean could stimulate photosynthesis in phytoplankton. The phytoplankton would convert the ocean's dissolved carbon dioxide into carbohydrate and oxygen gas, some of which would sink into the deeper ocean before oxidizing. More than a dozen open-sea experiments confirmed that adding iron to the ocean increases photosynthesis in phyto-plankton by up to 30 times. While this approach has been proposed as a potential solution to the ocean acidification problem, mitigation of surface ocean acidification might increase acidification in the less-inhabited deep ocean.

A report by the UK's Royal Society (2009) reviewed the approach for effectiveness, affordability, timeliness and safety. The rating for affordability was "medium", or "not expected to be very cost-effective." For the other three criteria, the ratings ranged from "low" to "very low" (i.e., not good). For example, in regards to safety, the report found a "[high] potential for undesirable ecological side effects," and that ocean fertilization "may increase anoxic regions of ocean ('dead zones')."

Carbon Negative Fuels

Carbonic acid can be extracted from seawater as carbon dioxide for use in making synthetic fuel. If the resulting flue exhaust gas was subject to carbon capture, then the process would be carbon negative over time, resulting in permanent extraction of inorganic carbon from seawater and the atmosphere with which seawater is in equilibrium. Based on the energy requirements, this process was estimated to cost about $50 per tonne of CO_2.

Effects of Global Warming on Marine Mammals

The effects of global warming on marine mammals are of growing concern. The effects can be both direct; like loss of habitat, temperature stress, and exposure to severe weather, and indirect; like changes in host pathogen associations, changes in body condition because of predator–prey interaction, changed in exposure to toxins, and increased human interactions. All of these are effects that are predicted that the marine mammals are living with because of the rapid climate change. The increase of carbon dioxide and other greenhouse gases into the atmosphere is thought to be the main cause of climate change or global warming. Exactly how this will affect the ocean, which is home to marine mammals, is hard to predict since there are many factors, such as weather events and salinity that affect ocean ecosystems. How all these will interact is highly unpredictable. Using global climate models (GCM)s, scientists can get a general idea of how climate change will impact the ocean environment in the future.

Effects

Marine mammals have evolved to live in the ocean, but the effects of climate change are rapidly altering their habitat. The climate change for arctic marine mammals by studying the difference in their habitat, distribution, abundance, movement and migration, body conditions, behavior, and their sensitivity. The climate change is drastically changing, so it is hard for some arctic marine mammals to adapt to the new conditions as quickly as needed, so the arctic marine mammals are being studied to see what happens to them due to the high stress they are put in due to the rapidly changing climate change. These climate changes put the arctic marine mammals in immense pressure and stress, due to this many species who cannot adapt begin to slowly die off due to the extra stress. As levels of greenhouse gases in the atmosphere increase they trap heat which causes an overall warming of the planet. During the last century, the global average land and sea surface temperature has increased dramatically. Warming has reached down to about 700 meters and more (per a study which finds that 30% of the ocean warming, over the past decade, has occurred in the deeper oceans below 700 meters). Many marine mammal species require specific temperature ranges in which they must live. The warming of the ocean will cause changes in the species migration or area that they inhibit. The species that cannot relocate due to some type of barrier will be forced to adapt to the increasingly warming sea waters or they will face extinction. Many of the species ranges are being pushed further and further north as water temperatures increase and they will soon have nowhere else to go.

The rise in sea level affect the coastal habitat and the species that rely on it.T he rising sea levels are causing erosion of the beaches, so this then causes the turtles to have a hard time finding a spot to lay their eggs. The rising temperature are also causing problems for the turtles because the temperature of the sand determines the sex of the eggs, so many places are having a 90% production of females and if the temperature rises by just one degree, there could potentially be no more male turtles. The right whales have been on the endangered species list since the 1970s and the population keep declining because of the decline of zooplankton. Because of the climate change, the population of zooplankton is decreasing and since this is the right whale's main food source, the female whales are not able to bulk up for the calving process. The seals are affected by the climate change because of the rapid increase of the melting ice. The seals use the ice to rest between their searches for food. With the increase of melting ice, the seals could be in trouble. This habitat is often used by pinnipeds as haul-out sites. In order to combat rising sea levels in areas inhabited by humans, the construction of sea walls has been proposed; however, these walls may interfere with the migration routes of several marine mammal species. These routes can be very important to some marine mammal species for reaching feeding and breeding grounds.

Changes in the temperature ranges will also change the location of areas with high primary productivity. These areas are important to marine mammals because primary producers are the food source of marine mammal prey or the marine mammals prey themselves. Marine mammal distribution and abundance will be determined by the distribution and abundance of its prey. Migration of marine mammals may also be affected by the changes in primary productivity.

Increased glacier ice melt also impacts ocean circulation because, due to the increase of freshwater in the ocean, salinity concentrations in the ocean are changing. Thermohaline circulation may be altered by increasing amounts of freshwater in the ocean. Thermohaline circulation is responsible for bringing up cold, nutrient-rich water from the depths of the ocean, a process known as upwelling. This may affect regional temperatures and primary productivity.

Susceptibility to disease is also thought to increase while reproductive success may decrease with increasing ocean temperatures.

The world's oceans absorb a large amount of carbon dioxide from the atmosphere, which causes an increase in carbon dioxide concentrations and decreases its overall pH, increasing ocean acidification.

Extinction Risk from Global Warming

The extinction risk of global warming is the risk of species becoming extinct due to the effects of global warming.

The scientific consensus in the IPCC Fourth Assessment Report is that

"Anthropogenic warming could lead to some impacts that are abrupt or irreversible, depending upon the rate and magnitude of the climate change."

"There is medium confidence that approximately 20-30% of species assessed so far are likely to be at increased risk of extinction if increases in global average warming exceed 1.5-2.5 °C (relative to 1980-1999). As global average temperature increase exceeds about 3.5 °C, model projections suggest significant extinctions (40-70% of species assessed) around the globe."

In one study published in *Nature* in 2004, between 15 and 37% of 1103 endemic or near-endemic known plant and animal species will be "committed to extinction" by 2050. More properly, changes in habitat by 2050 will put them outside the survival range for the inhabitants, thus committing the species to extinction.

Other researchers, such as Thuiller *et al.*, Araújo *et al.* , Person *et al.*, Buckley and Roughgarden, and Harte *et al.* have raised concern regarding uncertainty in Thomas *et al.*'s projections; some of these studies believe it is an overestimate, others believe the risk could be greater. Thomas *et al.* replied in Nature addressing criticisms and concluding "Although further investigation is needed into each of these areas, it is unlikely to result in substantially reduced estimates of extinction. Anthropogenic climate change seems set to generate very large numbers of species-level extinctions." On the other hand, Daniel Botkin *et al.* state "... global estimates of extinctions due to climate change (Thomas et al. 2004) may have greatly overestimated the probability of extinction..."

Mechanistic studies are documenting extinctions due to recent climate change: McLaughlin *et al.* documented two populations of Bay checkerspot butterfly being threatened by precipitation change. Parmesan states, "Few studies have been conducted at a scale that encompasses an entire species" and McLaughlin *et al.* agreed "few mechanistic studies have linked extinctions to recent climate change."

In 2008, the white lemuroid possum was reported to be the first known mammal species to be driven extinct by man-made global warming. However, these reports were based on a misunderstanding. One population of these possums in the mountain forests of northern Queensland is severely threatened by climate change as the animals cannot survive extended temperatures over 30 °C. However, another population 100 kilometres south remains in good health.

According to research published in the January 4, 2012 *Proceedings of the Royal Society B* current climate models may be flawed because they overlook two important factors: the differences in how quickly species relocate and competition among species. According to the researchers, led by Mark C. Urban, an ecologist at the University of Connecticut, diversity decreased when they took these factors into account, and that new communities of organisms, which do not exist today, emerged. As a result the rate of extinctions may be higher than previously projected.

According to research published in the 30 May 2014 issue of *Science,* most known species have small ranges, and the numbers of small-ranged species are increasing quickly. They are geographically concentrated and are disproportionately likely to be threatened or already extinct. According to the research, current rates of extinction are three orders of magnitude higher than the background extinction rate, and future rates, which depend on many factors, are poised to increase. Although there has been rapid progress in developing protected areas, such efforts are not ecologically representative, nor do they optimally protect biodiversity. In the researchers' view, human activity tends to destroy critical habitats where species live, warms the planet, and tends to move species around the planet to places where they don't belong and where they can come into conflict with human needs (e.g. causing species to become pests).

In 2016 the Bramble Cay melomys, which lived on a Great Barrier Reef island, was reported to probably be the first mammal to become extinct because of sea level rises due to human-made climate change.

Regional Effects of Global Warming

1999-2008 Mean Temperatures

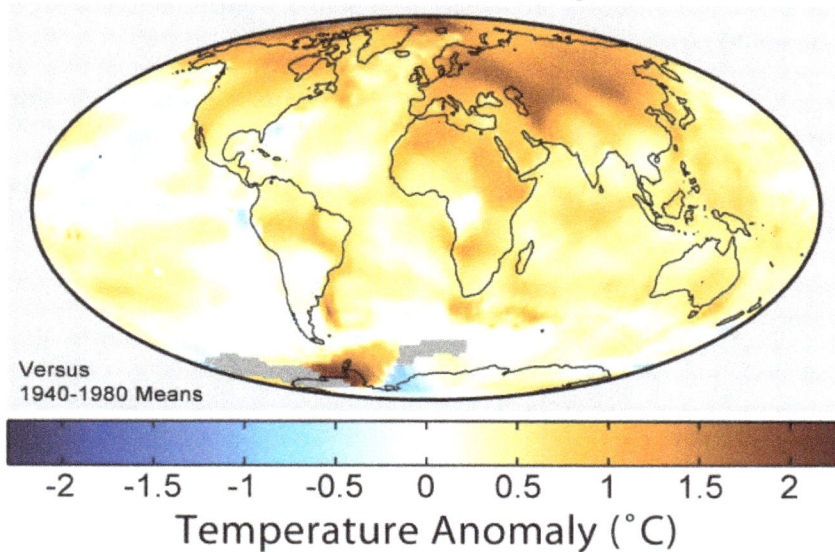

Versus
1940-1980 Means

Temperature Anomaly (°C)

Mean surface temperature change for 1999–2008 relative to the average temperatures from 1940 to 1980

Projected changes in average temperatures across the world in the 2050s under three greenhouse gas (GHG) emissions scenarios, relative to average temperatures between 1971-1999

Regional effects of global warming are long-term significant changes in the expected patterns of average weather of a specific region due to global warming. The world average temperature is rising due to the greenhouse effect caused by increasing levels of greenhouse gases, especially carbon dioxide. When the global temperature changes, the changes in climate are not expected to be uniform across the Earth. In particular, land areas change more quickly than oceans, and northern high latitudes change more quickly than the tropics, and the margins of biome regions change faster than do their cores.

Regional effects of global warming vary in nature. Some are the result of a generalised global change, such as rising temperature, resulting in local effects, such as melting ice. In other cases, a change may be related to a change in a particular ocean current or weather system. In such cases, the regional effect may be disproportionate and will not necessarily follow the global trend.

There are three major ways in which global warming will make changes to regional climate: melting or forming ice, changing the hydrological cycle (of evaporation and precipitation) and changing currents in the oceans and air flows in the atmosphere. The coast can also be considered a region, and will suffer severe impacts from sea level rise.

Regional impacts

Highlights of recent and projected regional impacts are shown below:

Impacts on Africa

- Africa is one of the most vulnerable continents to climate variability and change because of multiple existing stresses and low adaptive capacity. Existing stresses include poverty, political conflicts, and ecosystem degradation.

- By 2050, between 350 million and 600 million people are projected to experience increased water stress due to climate change

- Climate variability and change is projected to severely compromise agricultural production, including access to food, across Africa

- Toward the end of the 21st century, projected sea level rise will likely affect low-lying coastal areas with large populations

- Climate variability and change can negatively impact human health. In many African countries, other factors already threaten human health. For example, malaria threatens health in southern Africa and the Eastern Highlands.

Impacts on Arctic and Antarctic

- Climate change in the Arctic will likely reduce the thickness and extent of glaciers and ice sheets.

- Changes in natural ecosystems will likely have detrimental effects on many organisms including migratory birds, mammals, and higher predators. Climate change will likely cause changes in dominance structures in plant communities, with shrubs expanding

- In the Arctic, climate changes will likely reduce the extent of sea ice and permafrost, which can have mixed effects on human settlements. Negative impacts could include damage to infrastructure and changes to winter activities such as ice fishing and ice road transportation. Positive impacts could include more navigable northern sea routes.

- The reduction and melting of permafrost, sea level rise, and stronger storms may worsen coastal erosion.

- Terrestrial and marine ecosystems and habitats are projected to be at risk to invasive species, as climatic barriers are lowered in both polar regions.

Impacts on Asia

- Glaciers in Asia are melting at a faster rate than ever documented in historical records. Melting glaciers increase the risks of flooding and rock avalanches from destabilized slopes.

- Climate change is projected to decrease freshwater availability in central, south, east and southeast Asia, particularly in large river basins. With population growth and increasing demand from higher standards of living, this decrease could adversely affect more than a billion people by the 2050s.

- Increased flooding from the sea and, in some cases, from rivers, threatens coastal areas, especially heavily populated delta regions in south, east, and southeast Asia.

- By the mid-21st century, crop yields could increase up to 20% in east and southeast Asia. In the same period, yields could decrease up to 30% in central and south Asia.

- Sickness and death due to diarrhoeal disease are projected to increase in east, south, and southeast Asia due to projected changes in the hydrological cycle associated with climate change.

Impacts on Europe

- Wide-ranging impacts of climate change have already been documented in Europe. These impacts include retreating glaciers, longer growing seasons, species range shifts, and heat wave-related health impacts.

- Future impacts of climate change are projected to negatively affect nearly all European regions. Many economic sectors, such as agriculture and energy, could face challenges.

- In southern Europe, higher temperatures and drought may reduce water availability, hydropower potential, summer tourism, and crop productivity.

- In central and eastern Europe, summer precipitation is projected to decrease, causing higher water stress. Forest productivity is projected to decline. The frequency of peatland fires is projected to increase.

- In northern Europe, climate change is initially projected to bring mixed effects, including some benefits such as reduced demand for heating, increased crop yields, and increased forest growth. However, as climate change continues, negative impacts are likely to outweigh benefits. These include more frequent winter floods, endangered ecosystems, and increasing ground instability.

Impacts on Latin America

- By mid-century, increases in temperature and decreases in soil moisture are projected to cause savanna to gradually replace tropical forest in eastern Amazonia.

- In drier areas, climate change will likely worsen drought, leading to salinization (increased salt content) and desertification (land degradation) of agricultural land. The productivity of livestock and some important crops such as maize and coffee is projected to decrease, with adverse consequences for food security. In temperate zones, soybean yields are projected to increase.

- Sea level rise is projected to increase risk of flooding, displacement of people, salinization of drinking water resources, and coastal erosion in low-lying areas.

- Changes in precipitation patterns and the melting of glaciers are projected to significantly affect water availability for human consumption, agriculture, and energy generation.

Impacts on North America

Projected change in seasonal mean surface air temperature from the late 20th century (1971-2000 average) to the middle 21st century (2051-2060). The left panel shows changes for June–July–August (JJA) seasonal averages, and the right panel shows changes for December–January–February (DJF). The change is in response to increasing atmospheric concentrations of greenhouse gases and aerosols based on a "middle of the road" estimate of future emissions (SRES emissions scenario A1B). Warming is projected to be larger over continents than oceans, and is largest at high latitudes of the Northern Hemisphere during Northern Hemisphere winter (DJF) (Credit: NOAA Geophysical Fluid Dynamics Laboratory).

- Warming in western mountains is projected to decrease snowpack, increase winter flooding, and reduce summer flows, exacerbating competition for over-allocated water resources.

- Disturbances from pests, diseases, and fire are projected to increasingly affect forests, with extended periods of high fire risk and large increases in area burned.

- Moderate climate change in the early decades of the century is projected to increase aggregate yields of rain-fed agriculture by 5-20%, but with important variability among regions. Crops that are near the warm end of their suitable range or that depend on highly utilized water resources will likely face major challenges.

- Increases in the number, intensity, and duration of heat waves during the course of the century are projected to further challenge cities that currently experience heat waves, with potential for adverse health impacts. Older populations are most at risk.

- Climate change will likely increasingly stress coastal communities and habitats, worsening the existing stresses of development and pollution.

Impacts on Oceania

- Water security problems are projected to intensify by 2030 in southern and eastern Australia, and in the northern and some eastern parts of New Zealand.

- Significant loss of biodiversity is projected to occur by 2020 in some ecologically rich sites, including the Great Barrier Reef and the Wet Tropics of Queensland.

- Sea level rise and more severe storms and coastal flooding will likely impact coastal areas. Coastal development and population growth in areas such as Cairns and Southeast Queensland (Australia) and Northland to Bay of Plenty (New Zealand), would place more people and infrastructure at risk.

- By 2030, increased drought and fire is projected to cause declines in agricultural and forestry production over much of southern and eastern Australia and parts of eastern New Zealand.

- Extreme storm events are likely to increase failure of floodplain protection and urban drainage and sewerage, as well as damage from storms and fires.

- More heat waves may cause more deaths and more electrical blackouts.

Impacts on Small Islands

- Small islands, whether located in the tropics or higher latitudes, are already exposed to extreme weather events and changes in sea level. This existing exposure will likely make these areas sensitive to the effects of climate change.

- Deterioration in coastal conditions, such as beach erosion and coral bleaching, will likely affect local resources such as fisheries, as well as the value of tourism destinations.

- Sea level rise is projected to worsen inundation, storm surge, erosion, and other coastal hazards. These impacts would threaten vital infrastructure, settlements, and facilities that support the livelihood of island communities.

- By mid-century, on many small islands (such as the Caribbean and Pacific), climate change is projected to reduce already limited water resources to the point that they become insufficient to meet demand during low-rainfall periods.

- Invasion by non-native species is projected to increase with higher temperatures, particularly in mid- and high-latitude islands.

Inundation, Displacement, and National Sovereignty of Small Islands

According to scholar Tsosie, environmental disparities among disadvantaged communities including poor and racial minorities, extend to global inequalities between the developed and developing countries. For example, according to Barnett, J. and Adger, W.N. the projected damage to small islands and atoll communities will be a consequence of climate change caused by developing countries that will disproportionately affect these developing nations.

Sea-level rise and increased tropical cyclones are expected to place low-lying small islands in the Pacific, Indian, and Caribbean regions at risk of inundation and population displacement.

According to N. Mimura's study on the vulnerability of island countries in the South Pacific to sea level rise and climate change, financially burdened island populations living in the lowest-lying regions are most vulnerable to risks of inundation and displacement. On the islands of Fiji, Tonga and western Samoa for example, high concentrations of migrants that have moved from outer islands inhabit low and unsafe areas along the coasts.

Atoll nations, which include countries that are composed entirely of the smallest form of islands, called motus, are at risk of entire population displacement. These nations include Kiribati, Maldives, the Marshall Islands, Tokelau, and Tuvalu. According to a study on climate dangers to atoll countries, characteristics of atoll islands that make them vulnerable to sea level rise and other climate change impacts include their small size, their isolation from other land, their low income resources, and their lack of protective infrastructure.

A study that engaged the experiences of residents in atoll communities found that the cultural identities of these populations are strongly tied to these lands. The risk of losing these lands therefore threatens the national sovereignty, or right to self-determination, of Atoll nations.

Human rights activists argue that the potential loss of entire atoll countries, and consequently the loss of cultures and indigenous lifeways cannot be compensated with financial means. Some researchers suggest that the focus of international dialogues on these issues should shift from ways to relocate entire communities to strategies that instead allow for these communities to remain on their lands.

Especially Affected Regions

The Arctic, Africa, small islands and Asian megadeltas are regions that are likely to be especially affected by future climate change. Within other areas, some people are particularly at risk from future climate change, such as the poor, young children and the elderly.

The Arctic

The Arctic is likely to be especially affected by climate change because of the high projected rate of regional warming and associated impacts. Temperature projections for the Arctic region were assessed by Anisimov *et al.* (2007). These suggested areally averaged warming of about 2 °C to 9 °C by the year 2100. The range reflects different projections made by different climate models, run with different forcing scenarios. Radiative forcing is a measure of the effect of natural and human activities on the climate. Different forcing scenarios reflect, for example, different projections of future human greenhouse gas emissions.

Africa

Africa is likely to be the continent most vulnerable to climate change. With high confidence, Boko *et al.* (2007) projected that in many African countries and regions, agricultural production and food security would probably be severely compromised by climate change and climate variability.

The United Nations Environment Programme (UNEP, 2007) produced a post-conflict environmental assessment of Sudan. According to UNEP (2007), environmental stresses in Sudan are interlinked with other social, economic and political issues, such as population displacement and competition over natural resources. Regional climate change, through decreased precipitation, was thought to have been one of the factors which contributed to the conflict in Darfur. Along with other environmental issues, climate change could negatively affect future development in Sudan. One of the recommendations made by UNEP (2007) was for the international community to assist Sudan in adapting to climate change.

Small Islands

On small islands, sea level rise is expected to exacerbate inundation, erosion and other coastal hazards, and threaten vital infrastructure, human settlements and facilities that support the livelihood of island communities. In the coastal zone of Asia, there are 11 megadeltas with an area greater than 10,000 km^2. These megadeltas are homes to millions of people, and contain diverse ecosystems. Climate change and sea level rise could increase the frequency and level of inundation of Asian megadeltas due to storm surges and floods from river drainage.

Ice-cover Changes

Permanent ice cover on land is a result of a combination of low peak temperatures and sufficient precipitation. Some of the coldest places on Earth, such as the dry valleys of Antarctica, lack significant ice or snow coverage due to a lack of snow. Sea ice however maybe formed simply by low temperature, although precipitation may influence its stability by changing albedo, providing an insulating covering of snow and affecting heat transfer. Global warming has the capacity to alter both precipitation and temperature, resulting in significant changes to ice cover. Furthermore, the behaviour of ice sheets, ice caps and glaciers is altered by changes in temperature and precipitation, particularly as regards the behaviour of water flowing into and through the ice.

Arctic Sea Ice

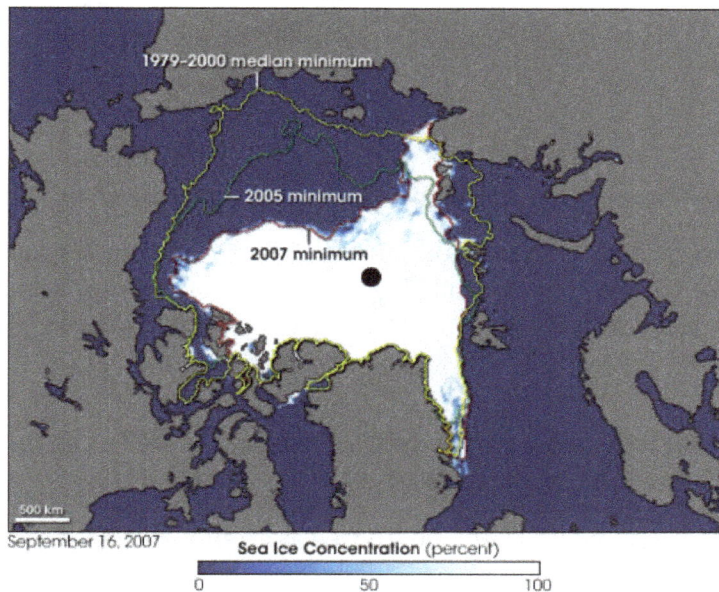

Arctic sea ice minima in 2005, 2007, and the 1979-2000 average.

Recent projections of sea ice loss suggest that the Arctic ocean will likely be free of summer sea ice sometime between 2059 and 2078.

Models showing decreasing sea ice also show a corresponding decrease in polar bear habitat. Some scientists see the polar bear as a species which will be affected first and most severely by global warming because it is a top-level predator in the Arctic, which is projected to warm more than the global average. Recent reports show polar bears resorting to cannibalism, and scientists state that these are the only instances that they have observed of polar bears stalking and killing one another for food.

Antarctica

The Antarctic peninsula has lost a number of ice shelves recently. These are large areas of floating ice which are fed by glaciers. Many are the size of a small country. The sudden collapse of the Larsen B ice shelf in 2002 took 5 weeks or less and may have been due to global warming. Larsen B had previously been stable for up to 12,000 years.

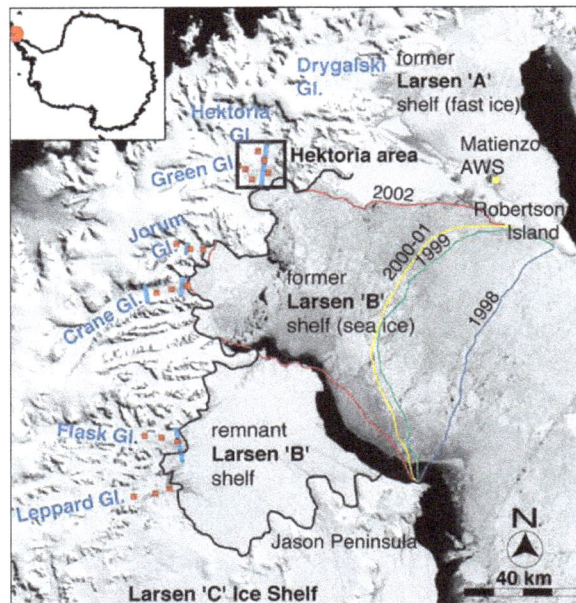

The collapse of Larsen B, showing the diminishing extent of the shelf from 1998 to 2002

Concern has been expressed about the stability of the West Antarctic ice sheet. A collapse of the West Antarctic ice sheet could occur "within 300 years [as] a worst-case scenario. Rapid sea-level rise (>1 m per century) is more likely to come from the WAIS than from the [Greenland ice sheet]."

Greenland

As the Greenland ice sheet loses mass from calving of icebergs as well as by melting of ice, any such processes tend to accelerate the loss of the ice sheet.

The IPCC suggest that Greenland will become ice free at around 5 Celsius degrees over pre-industrial levels, but subsequent research comparing data from the Eemian period suggests that the ice sheet will remain at least in part at these temperatures. The volume of ice in the Greenland sheet is sufficient to cause a global sea level rise of 7 meters. It would take 3,000 years to completely melt the Greenland ice sheet. This figure was derived from the assumed levels of greenhouse gases over the duration of the experiment. In reality, these greenhouse gas levels are of course affected by future emissions and may differ from the assumptions made in the model.

Glaciers

Glacier retreat not only affects the communities and ecosystems around the actual glacier, but the entire downstream region. The most notable example of this is in India, where river systems such as the Indus and Ganges are ultimately fed by glacial meltwater from the Himalayas. Loss of these glaciers will have dramatic effects on the downstream region, increasing the risk of drought as lower flows of meltwater reduce summer river flows unless summer precipitation increases. Altered patterns of flooding can also affect soil fertility.

The Tibetan Plateau contains the world's third-largest store of ice. Qin Dahe, the former head of the China Meteorological Administration, said that the recent fast pace of melting and warmer temperatures will be good for agriculture and tourism in the short term; but issued a strong warning:

"Temperatures are rising four times faster than elsewhere in China, and the Tibetan glaciers are retreating at a higher speed than in any other part of the world.... In the short term, this will cause lakes to expand and bring floods and mudflows. . . . In the long run, the glaciers are vital lifelines for Asian rivers, including the Indus and the Ganges. Once they vanish, water supplies in those regions will be in peril."

Permafrost Regions

Regions of permafrost cover much of the Arctic. In many areas, permafrost is melting, leading to the formation of a boggy, undulating landscape filled with thermokarst lakes and distinctive patterns of drunken trees. The process of permafrost melting is complex and poorly understood since existing models do not include feedback effects such as the heat generated by decomposition.

Arctic permafrost soils are estimated to store twice as much carbon as is currently present in the atmosphere in the form of CO_2. Warming in the Arctic is causing increased emissions of CO_2 and Methane (CH_4).

Precipitation and Vegetation Changes

The Eastern Amazon rainforest may be replaced by Caatinga vegetation as a result of global warming.

Much of the effect of global warming is felt through its influence on rain and snow. Regions may become wetter, drier, or may experience changes in the intensity of precipitation - such as moving from a damp climate to one defined by a mixture of floods and droughts. These changes may have a very severe impact on both the natural world and human civilisation, as both naturally occurring and farmed plants experience regional climate change that is beyond their ability to tolerate.

A U.S. National Oceanic and Atmospheric Administration (NOAA) analysis published in the Journal of Climate October 2011, and cited on Joseph J. Romm's, climateprogress.org, found that increasing droughts in the Middle East during the wintertime when the region traditionally most of its rainfall to replenish aquifers, and anthropogenic climate change is partly responsible. Per Earth System Research Laboratory's Martin Hoerling "The magnitude and frequency of the drying that has occurred is too great to be explained by natural variability alone," and "This is not encouraging news for a region that already experiences water stress, because it implies natural variability alone is unlikely to return the region's climate to normal." the lead author of the paper. Twelve of the world's fifteen most water-scarce countries — Bahrain, Qatar, Algeria, Libya, Tunisia, Jordan, Saudi Arabia, Yemen, Oman, the United Arab Emirates, Kuwait, Israel and Palestine — are in the Middle East.

Arctic and Alpine Regions

Polar and alpine ecosystems are assumed to be particularly vulnerable to climate change as their organisms dwell at temperatures just above the zero degree threshold for a very short summer growing season. Predicted changes in climate over the next 100 years are expected to be substantial in arctic and sub-arctic regions. Already there is evidence of upward shifts of plants in mountains and in arctic shrubs are predicted to increase substantially to warming

Amazon

One modeling study suggested that the extent of the Amazon rainforest may be reduced by 70% if global warming continues unchecked, due to regional precipitation changes that result from weakening of large-scale tropical circulation.

Sahara

Some studies suggest that the Sahara desert may have been more vegetated during the warmer Mid-Holocene period, and that future warming may result in similar patterns.

Sahel

Some studies have found a greening of the Sahel due to global warming. Other climate models predict "a doubling of the number of anomalously dry years [in the Sahel] by the end of the century".

Desert Expansion

Expansion of subtropical deserts is expected as a result of global warming, due to expansion of the Hadley Cell.

Coastal Regions

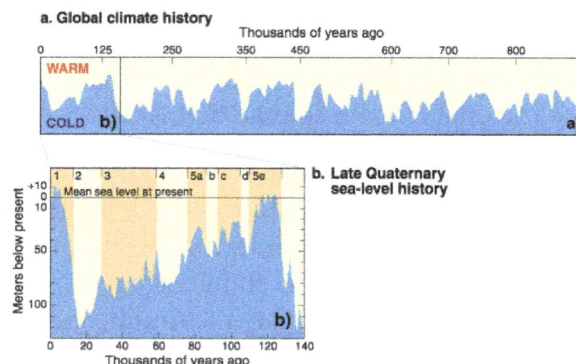

Past sea-level changes and relative temperatures. Global warming is expected to dramatically affect sea level.

Global sea level is currently rising due to the thermal expansion of water in the oceans and the addition of water from ice sheets. Because of this, there low-lying coastal areas, many of which are heavily populated, are at risk of flooding.

Areas threatened by current sea level rise include Tuvalu and the Maldives. Regions that are prone to storm surges, such as London, are also threatened.

With very high confidence, IPCC (2007) projected that by the 2080s, many millions more people would experience floods every year due to sea level rise. The numbers affected were projected to be largest in the densely populated and low-lying megadeltas of Asia and Africa. Small islands were judged to be especially vulnerable.

Ocean Effects

North Atlantic Region

It has been suggested that a shutdown of the Atlantic thermohaline circulation may result in relative cooling of the North Atlantic region by up to 8C in certain locations. Recent research suggests that this process is not currently underway.

Tropical Surface and Troposphere Temperatures

In the tropics, basic physical considerations, climate models, and multiple independent data sets indicate that the warming trend due to well-mixed greenhouse gases should be faster in the troposphere than at the surface.

References

- Committee on Ecological Impacts of Climate Change, US National Research Council (NRC) (2008). Ecological Impacts of Climate Change. 500 Fifth Street, NW Washington, DC 20001, USA: The National Academies Press. ISBN 978-0-309-12710-3.

- Karl, Thomas R.; Melillo, Jerry M.; Peterson, Thomas C., eds. (2009). Global Climate Change Impacts in the United States (PDF). New York: Cambridge University Press. ISBN 978-0-521-14407-0.

- US NRC (2010). Advancing the Science of Climate Change. A report by the US National Research Council (NRC). Washington, D.C., USA: National Academies Press. ISBN 0-309-14588-0.

- Kump, Lee R.; Kasting, James F.; Crane, Robert G. (2003). The Earth System (2nd ed.). Upper Saddle River: Prentice Hall. pp. 162–164. ISBN 0-613-91814-2.

- UK Royal Society (September 2009). "Geoengineering the climate: science, governance and uncertainty" (PDF). London: UK Royal Society. ISBN 978-0-85403-773-5.

- UNEP (November 2010). "The Emissions Gap Report: Are the Copenhagen Accord pledges sufficient to limit global warming to 2°C or 1.5°C? A preliminary assessment". Nairobi, Kenya: United Nations Environment Programme (UNEP). ISBN 978-92-807-3134-7

- US NRC (2011). America's Climate Choices. A report by the Committee on America's Climate Choices, US National Research Council (US NRC). Washington, DC, USA: National Academies Press. ISBN 978-0-309-14585-5

- WBGU (2006). Special Report: The Future Oceans – Warming Up, Rising High, Turning Sour (PDF). Berlin, Germany: WBGU. ISBN 3-936191-14-X.

- Roxy, M.K. (2016). "A reduction in marine primary productivity driven by rapid warming over the tropical Indian Ocean". Geophysical Research Letters. 43 (2). doi:10.1002/2015GL066979. Retrieved January 2016.

- Kump, L.R.; Bralower, T.J.; Ridgwell, A. (2009). "Ocean acidification in deep time". Oceanography. 22: 94–107. doi:10.5670/oceanog.2009.10. Retrieved 16 May 2016.

- Zeebe, R.E. (2012). "History of Seawater Carbonate Chemistry, Atmospheric CO2, and Ocean Acidification". Annual Review of Earth and Planetary Sciences. 40: 141–165. doi:10.1146/annurev-earth-042711-105521. Retrieved 16 May 2016.

- "Feely et al. - Evidence for upwelling of corrosive "acidified" water onto the Continental Shel". pmel.noaa.gov. Retrieved 2014-01-25.

- "An Ominous Warning on the Effects of Ocean Acidification by Carl Zimmer: Yale Environment 360". e360.yale.edu. Retrieved 2014-01-25.

Permissions

Index

www.ingramcontent.com/pod-product-compliance
Lightning Source LLC
Chambersburg PA
CBHW061243190326
41458CB00011B/3561